W9-CMM-124

ground water

Protection Alternatives and Strategies in the U.S.A.

Compiled by members of the Task Committee
on Ground Water Protection of the
Ground Water Hydrology Committee of the Hydrology
Technical Oversight Committee of the
Water Resources Engineering Division of the
American Society of Civil Engineers

Nazeer Ahmed, Chairman

Robert McVicker, Secretary

Thomas W. Anderson

Carl Carpenter

Wendy L. Cohen

Marvin Damm

A. Ivan Johnson

Richard Peralta

J. Paul Riley

Published by

ASCE *American Society of Civil Engineers*

345 East 47th Street
New York, New York 10017-2398

TECHNICAL COLLEGE OF THE LOWCOUNTRY
LEARNING RESOURCES CENTER
POST OFFICE BOX 1288
BEAUFORT, SOUTH CAROLINA 29901-1288

Abstract:

This report of the Task Committee on Ground Water Protection of the Water Resources Engineering Division of the American Society of Civil Engineers presents general hydrogeologic information on the occurrence of contamination in ground water that applies nationwide, and specific examples of approaches to ground water protection used in various regions of the United States. Since management and regulation of resources by regional authorities is one of the most important approaches to ground water protection, the majority of the report discusses the various approaches tried by different state and local authorities. The great variety of strategies adopted by these agencies indicate the uniqueness of each aquifer and the need to adjust any strategy to account for these site specific factors.

Library of Congress Cataloging-in-Publication Data
Ground water protection alternatives and strategies in the U.S.A. / compiled by members of the Task Committee on Ground Water Protection of the Ground Water Hydrology Committee of the Hydrology Technical Oversight Committee of the Water Resources Engineering Division of the American Society of Civil Engineers : Nazeer Ahmed, chairman ; Thomas W. Anderson ... [et al.].
    p.   cm.
ISBN 0-7844-0231-0
1. Groundwater--Pollution--United States. 2. Wellhead protection--United States. I. Ahmed, Nazeer, 1935-   . II. American Society of Civil Engineers. Task Committee on Ground Water Protection.
TD223.G736 1997   97-8835
363.739'47'0973--dc21   CIP

        The material presented in this publication has been prepared in accordance with generally recognized engineering principles and practices, and is for general information only. This information should not be used without first securing competent advice with respect to its suitability for any general or specific application.
        The contents of this publication are not intended to be and should not be construed to be a standard of the American Society of Civil Engineers (ASCE) and are not intended for use as a reference in purchase specifications, contracts, regulations, statutes, or any other legal document.
        No reference made in this publication to any specific method, product, process or service constitutes or implies an endorsement, recommendation, or warranty thereof by ASCE.
        ASCE makes no representation or warranty of any kind, whether express or implied, concerning the accuracy, completeness, suitability, or utility of any information, apparatus, product, or process discussed in this publication, and assumes no liability therefore.
        Anyone utilizing this information assumes all liability arising from such use, including but not limited to infringement of any patent or patents.

Photocopies. Authorization to photocopy material for internal or personal use under circumstances not falling within the fair use provisions of the Copyright Act is granted by ASCE to libraries and other users registered with the Copyright Clearance Center (CCC) Transactional Reporting Service, provided that the base fee of $4.00 per article plus $.25 per page is paid directly to CCC, 222 Rosewood Drive, Danvers, MA 01923. The identification for ASCE Books is 0-7844-0231-0/97/$4.00 + $.25 per page. Requests for special permission or bulk copying should be addressed to Permissions & Copyright Dept., ASCE.

Copyright © 1997 by the American Society of Civil Engineers,
All Rights Reserved.
Library of Congress Catalog Card No: 97-8835
ISBN 0-7844-0231-0
Manufactured in the United States of America.

TABLE OF CONTENTS

INTRODUCTION

T.W. ANDERSON[1], M. ASCE

BACKGROUND

In 1986, the Task Committee on Ground Water Protection Alternatives was formed under the parent Ground Water Committee of the ASCE Irrigation and Drainage Division. The purposes of the new Task Committee were to review ground water contamination problems around the United States and to inventory and evaluate alternatives for the protection of ground water. The Task Committee organized technical sessions for Irrigation and Drainage Division Specialty Conferences from 1987 through 1992, with the goal of presenting information and discussion of various approaches utilized throughout the United States to protect ground water quantity and quality. An initial product of this Task Committee was a lay-person brochure, which presents information on basic ground water hydrology, summarizes mechanisms through which contamination of ground water could occur, and discusses alternatives for protection of ground water.

In 1992, after a 6-year period of highly successful technical sessions at Irrigation and Drainage Division Specialty Conferences at various locations around the United States, a follow-up Task Committee was formed to bring together, update, and publish a

---

[1] Senior Hydrologist, Errol L. Montgomery & Associates, Inc., 1550 E. Prince Road, Tucson, Arizona, 85719, Phone (520) 881-4912

1

compilation of the important papers presented at those Specialty Conferences on strategies to protect ground water around the country. This report is the product of that Task Committee. The purpose of this report is to present general hydrogeologic information on the occurrence of contamination in ground water that would apply nationwide, and specific examples of approaches to protection of ground water used in various regions of the United States.

IDENTIFICATION OF THE PROBLEM

Ground water is widely available throughout the U.S.A. and is one of the Nation's most valuable natural resources. Ground water use has increased dramatically in recent decades. In 1950, estimated ground water use was 34 billion gallons per day (bgd) and, in 1990, estimated ground water use was about 80 bgd (Solley and others, 1993). Ground water is estimated to provide the source of drinking water for more than 50 percent of the U.S. population. Consequently, the protection of quantity and quality of ground water is of importance to most citizens.

Protection of ground water quantity is important to avoid substantial depletion of the resource. Ground water has been extensively used for irrigation because of the ability to obtain the resource at or very near the point of use and because of the reliability of the supply. Extended periods of drought conditions do not substantially affect ground water supplies. Extensive wellfields for pumping ground water have been developed throughout many areas of the U.S.A., particularly for irrigation use in much of the midwest and west. Declines in ground water levels and associated depletion of aquifer storage have occurred due to pumping in excess of natural recharge in many developed areas. Adverse effects that are associated with decline in ground water levels include increased pumping cost, decreased well yield, decreased discharge to streams, and land subsidence in those areas where the geologic environment is subject to this effect.

Under natural conditions, the chemical quality of ground water is generally good. However, many actions by humans can cause contamination of ground water, with a result that part of the ground water resource becomes unfit for human consumption. Some of the potential sources of contamination include:

- surface application of herbicides, fungicides, insecticides, and fertilizers

- leachate from landfills and dumps

- leakage from underground storage tanks

- accidental spills and improper disposal of organic solvents

- leakage from surface impoundments containing community or industrial waste

- improperly abandoned wells and improperly constructed injection wells

- highway deicing salt

The time required for migration of contaminants through the unsaturated zone depends on the depth to ground water and the geologic nature of the rock unit. Because velocity of ground water movement typically is small, ground water contamination may not be detected for years after a land-use practice causing the source of contaminants has been discontinued.

There are several remediation techniques used at contaminated sites, including "pump-and-treat", in-situ bioremediation, and soil vapor extraction. The first technique involves pumping contaminated ground water and treating it to remove contaminants. In-situ bioremediation enhances existing populations of microbes which degrade the contaminants in soil and ground water by using them as a food source. With soil vapor extraction, contaminated vapors are removed from the soil and treated before discharge to the ambient air. All of these techniques, as well as other less commonly used techniques, require an extended period of time to effect clean-up. More effective and efficient clean-up techniques are needed. However, protection of ground water from the occurrence of contamination is a more favorable alternative.

GROUND WATER PROTECTION

Protection of ground water from extreme depletion and contamination involves several factors. In order to protect ground water, it is necessary to understand the basin principles of occurrence and movement of ground water in an aquifer system. Public education is an important aspect of ground water protection. Monitoring the quantity and quality of ground water is also important for protection of ground water resources. Because of the nature of ground water systems and the fact that aquifer boundaries do not coincide with political boundaries, it is often necessary to involve multiple levels of government in efforts to protect ground water resources.

Areas in which recharge to the ground water system occurs are particularly vulnerable to the occurrence of contamination. In addition, areas of shallow ground water can be readily contaminated by percolation from land surface. Because of the nature of the areas of potential contamination, land-use zoning and regulation are important means for protecting ground water. Other methods of protection include controls on the use of potentially harmful liquids, such as insecticides and solvents. Probably the most important approach to protection of ground water is management and regulation of resources by individual States or local authorities. Management and regulation are accomplished by establishment of State or local rules on land use and chemical use, and by establishment of effluent standards and disposal rules. No particular approach is appropriate for all sources of contamination.

Discussions included in the papers contained in this report identify a number of approaches to protect ground water. The papers are grouped in six categories as follows: 1) General/National, which contains papers of generic applicability throughout the U.S.A.; 2) Northeast Region; 3) Mid-Atlantic Region; 4) Southeast Region; 5) Pacific-Southwest Region; and 6) Plains Region. The topics presented in the papers cover a broad range; State and local regulatory approaches to ground water protection are emphasized.

REFERENCES
Solley, W.B., Pierce, R.R., and Perlman, H.A., 1993, **Estimated use of water in the United States in 1990:** U.S. Geological Survey Circular 1081, 76 p.

Principles of Groundwater Protection

David W. Miller[1]

Introduction

Each year industry and government earmark hundreds of millions of dollars toward solving groundwater quality problems in response to such laws as the Resource Conservation and Recovery Act and Superfund. This effort is directed almost solely toward cleaning up existing groundwater contamination. What is needed today is a greater priority (with the corresponding increase in funding) for groundwater protection -- working to prevent groundwater pollution in the first place. Regulators are beginning to understand that once a portion of an aquifer has been severely contaminated it's usefulness as a source of drinking water is essentially eliminated forever. For this reason, programs such as wellhead protection, which was included in the reauthorization of the Safe Drinking Water Act, are essential.

Below is an introduction on how to implement and carry out wellhead protection programs and also some thoughts about groundwater. Basic principles -- ranging from the many sources of contamination to how groundwater occurs and how contaminants migrate -- are also discussed below.

Sources of Contamination

Federal and state laws designed to protect groundwater have focused on chemical contamination from landfills and surface impoundments at industrial facilities. However, landfills and wastewater impoundments are not the only sources of groundwater contamination. Other industry-related sources include chemical leaks from storage areas, accidental spills, vapor

---

[1]President & CEO, Geraghty & Miller, Inc., Environmental Services, 125 East Bethpage Road, Plainview, NY 11803

condensate from solvent recovery systems, and poor housekeeping. Common commercial operations, such as automotive repair or body shops, junkyards, dry cleaning outlets, and printing plants are often unsuspected contributors to local and regional groundwater contamination.

Non-industrial sources of groundwater pollutants include road runoff, municipal landfills, domestic wastewater, and farms with their feed lots and use of pesticides and fertilizers. Household products contain many organic chemicals that find their way into septic tanks, cesspools, and leaky sewer lines. From these sources contaminants may migrate to the water table.

<u>Groundwater Protection</u>

The recognition of these potential sources of groundwater contamination has resulted in the application of good housekeeping or best-management practices (BMP's) at many industrial sites. BMP's include: STORAGE - the containment of stored raw materials and waste products; INSPECTION - of processes that might lead to discharge to the ground and control over surface drainage; SPILL PREVENTION and RESPONSE; COMPLIANCE AUDITS; and long-term MONITORING.

Federal and state programs directed toward wellhead protection are also being developed based on the reauthorization of the Safe Drinking Water Act and local regulations. Under such programs, the capture zone contributing recharge to community water-supply wells is mapped using the results from pumping tests and water-level observations. The actual size of the capture zone is then computed either analytically or by means of a numerical model.

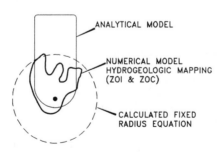

WHPA COMPARATIVE ANALYSIS EXAMPLE

Figure 1

Sources of contamination within each capture zone are regulated by state and local agencies through zoning, toxic substance control bylaws, or health-based restrictions. Figure 1 compares the capture zones around a particular well field using various methods of analysis. Wellhead protection zones defined for Dade County, FL, are shown in Figure 2. Here, land use controls are based on time of travel of groundwater to the individual well fields.

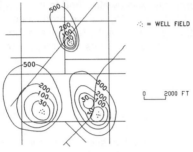

TRAVEL TIME TO WELL FIELDS
(in days)

Figure 2

Mechanisms Affecting Groundwater Contamination

Knowledge of the groundwater flow system is an essential precondition to developing a wellhead protection program. A monitoring program implemented without adequate hydrogeological information can be misleading. A body of contaminated groundwater may contain the accumulation of decades of leachate discharge, and it may take many years for contaminants to be detected in a nearby water-supply well. In addition to understanding flow, an appreciation of how chemicals move with groundwater is essential.

Chemicals may pass through several hydrologic zones as they migrate to the water table, including the agricultural soil zone, a deeper unsaturated soil zone, and the capillary fringe. The pore spaces in the agricultural zone are unsaturated, that is, occupied by both air and water. Flow in this zone is downward, as liquid contaminants or solutions of contaminants move under the force of gravity.

The agricultural soil zone is an important area for pollutant attenuation. Some chemicals are retained in this zone by adsorption onto organic material and chemically active soil particles. These adsorbed chemicals can subsequently be decomposed through such processes as oxidation and microbial activity. Many end-products of decomposition are taken up by plants or released to the atmosphere.

Additional soil below the agricultural zone may also have pore spaces that are unsaturated. As rainwater and snowmelt with chemicals percolate through this zone, chemical and biological degradation continue to take place. Some chemicals are adsorbed in this zone, and precipitation of inorganic solids may occur.

In the capillary fringe, spaces between soil particles may be saturated by water rising from the water table under capillary forces. Certain chemicals like hydrocarbons that are lighter than water "float" on top of the water table. These floating chemicals form a separate phase and may move in different directions and at different rates than contaminants that are dissolved in the percolating recharge. Once dissolved contaminants reach the water table, they enter the groundwater flow system, which has both horizontal and vertical components. Below the water table, all pore spaces between soil particles are saturated.

Unlike the turbulent flow of surface-water systems, the flow of groundwater is laminar; particles of fluid move along distinct and separate paths, with little mixing as the groundwater moves. Dissolved chemicals in the saturated zone flow with the groundwater; the direction of flow is governed by differences in hydrostatic head. Groundwater flow rates in aquifers generally range from a fraction of an inch to a few feet per day. The relative unavailability of dissolved oxygen in the saturated zone limits the potential for the oxidation of chemicals. Varying levels of attenuation may take place, depending on geologic conditions.

Plume Formation and Movement

Because groundwater flow is laminar, dissolved chemicals follow groundwater flow lines and form distinct plumes. Contaminated groundwater flows through the aquifer from areas of recharge to areas of discharge such as rivers and estuaries, as shown on Figure 3.

Concentrations of the dissolved chemicals are typically too low to affect the density of groundwater in the plume. Plumes of contaminated groundwater have been traced from a few feet to several miles downgradient of a pollution source. The shape and size of a plume depend on a number of factors, including the local geologic framework, local and regional groundwater flow, the type and concentration of contaminants, and variations in the rates of leaching.

Predicting the migration of contaminants through fractured rock or solution cavities is much more complex than predicting flow through sand aquifers. Contaminated groundwater moves through preferred and sometimes complex pathways.

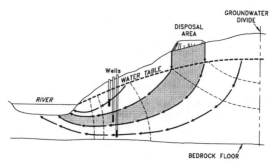

FLOW OF CONTAMINANTS IN A WATER—TABLE AQUIFER
(humid region)

Figure 3

Various chemicals are differentially attenuated in the soil through adsorption and chemical interaction with other organic and inorganic constituents of the aquifer to varying degrees.   Variables affecting contaminant mobility are water solubility and the potential for chemical degradation. Volatile organic chemicals in groundwater are fairly mobile; dispersion is the principal physical factor causing dilution. In-situ bacteria in the saturated zone can destroy and/or transform many organic chemicals. Because the contaminant chemicals also interact with each other, only estimates of their movement and fate in groundwater are possible.

The density of contaminated fluids is another factor in the formation and movement of a plume.   Slightly soluble materials may flow in separate phases. For example, oil can move as a body flowing on the surface of the groundwater table. In addition, the undissolved phase may give off vapors that migrate through the unsaturated zone in patterns unrelated to the groundwater flow system.

Surveys of groundwater contamination can be complicated by the variability of operating practices at typical waste disposal facilities. Numerous distinct plumes of contamination can move independently away from a site. Discontinuous discharges may result in "slugs" of contaminated water, causing wide spatial and temporal fluctuations in well-water quality. Lenses of sand and clay can cause other variations because they stratify the contaminants. Pumping from wells can modify groundwater flow patterns and, consequently, alter the movement of a contaminant plume.

Detailed monitoring of sites where groundwater contamination has been present for more than 5 years has revealed fluctuations in the concentration of some constituents while other constituents remained relatively constant.

This phenomenon is caused by the solution and dissolution of certain chemicals as the contaminant plume interacts with geologic materials in its path.

Dense Non-Aqueous Phase Liquids (DNAPLs)

Spills and leaks of dense chlorinated solvents can penetrate the unsaturated zone and flow into porous and fractured media in a non-aqueous phase. In some cases the volume is great enough for the chlorinated hydrocarbons to sink all the way to the bottom of the aquifer and to spread along the underlying confining layer. Whether the non-aqueous phase liquid reaches the bottom of the aquifer or not, some of the chlorinated solvents are retained in the porous or fractured media as globules. The solvent can spread out in more permeable zones overlying less permeable zones. Figure 4 shows a site where large volumes of both floating and sinking free product have been released to the aquifer. Non-aqueous phase liquids are found on the water table and at the bottom of the aquifer above the confining layer.

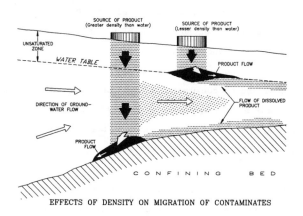

EFFECTS OF DENSITY ON MIGRATION OF CONTAMINATES

Figure 4

The presence of non-aqueous phase liquids extends the period of time needed to remove dissolved contaminants from an aquifer through pumping and treating the groundwater. Very limited success should be anticipated in removing the DNAPLs themselves.

Use of Analytical Models for
Ground Water Protection Design

Patrick G. Sloan, P.E. and Ajay Chandwani[1]

INTRODUCTION

As protection of ground water quality becomes paramount among environmental concerns, ground water flow and contaminant transport modeling are becoming increasingly used tools. As with any tool, the danger for misuse is present, especially when the tool is relatively new to the user and/or consumer. Some of the dangers in ground-water modeling include misconceptualization of the physical system, over-simplification, over-calibration, misapplication of the code, and unrealistic expectations.

The ASTM Ground Water Modeling Task Group has defined three types of ground water models; learning tools, engineering calculations, and aquifer simulators (Brown, 1992). It seems that many automatically think of aquifer simulators when ground water modeling is discussed.

Daily & Associates, Engineers, Inc.
7500 N. Harker Drive
Peoria, Illinois 61615

309/691/5300
309/691-1892 (FAX)

Maybe this is because they seem more glamorous, have been associated with many superfund sites, or because they can be more challenging for the users. In any case, the value of ground water flow and contaminant transport modeling by simpler models is often ignored or discounted.

The primary steps of ground water modeling are: 1) define study objectives, 2) develop a conceptual model, 3) select a computer code, 4) construct a ground water flow model, 5) calibrate model and perform sensitivity analysis, 6) make predictive simulations, 7) document model study, and 8) perform post audit. These steps are necessary for any modeling project. The type of model is determined in step 3, selection of a code. At this stage, the ground water modeling professional should give strong consideration to select the simplest code that is consistent with the objectives and conceptual model. The advantages include: 1) all remaining modeling steps become easier, 2) cost savings for the client, 3) danger of over-calibration or misapplication are less, 4) the answer is probably the same as what one would get from a much more expensive aquifer simulator, and 5) focus remains on the physical problem at hand and a design or remediation project that is protective of the environment. The disadvantages include: 1) consumers and regulators may need convincing that simpler does not mean less valid, and 2) some domain of the site may be oversimplified and the resulting design may be overly conservative. However, review of remediation and containment projects suggests that the danger of an overly conservative design is minimal, the cost savings in not using a complicated aquifer simulator may pay for conservative design, and ground water quality protection is not lessened.

A case study of the Peoria Landfill is presented below to illustrate the points discussed above. The landfill is located at an unreclaimed strip mine site in north-central Illinois. It was surface mined in the 1960's and landfilling began in 1975.

STUDY OBJECTIVES

Ground water contaminant transport modeling is required at new landfills or expansions in the State of Illinois. Federal Subtitle D regulations require modeling if a liner system other than a composite liner is proposed. The ground water impact of a landfill is acceptable in Illinois if the ground water contaminant transport (GCT) model predicts that concentrations of expected leachate constituents remain less than the background ground water quality for at least 100 years past closure and 30.5 m beyond the waste boundary. In other words, a mixing zone, called a zone of attenuation, is defined below the landfill and 30.5 m beyond the waste limit in which contamination is allowed in the uppermost aquifer. Beyond this zone no contamination is allowed.

CONCEPTUAL MODEL

Site Description

Section I of the Peoria Landfill was constructed without a liner (other than the in-situ material) and leachate collection system (Figure 1). Section II, which began operation in 1987, has the in-situ material plus a recompacted earth liner 0.61 m thick and a hydraulic conductivity, K, $\leq$ $1x10^{-7}$ cm/sec. Section II possesses a rudimentary leachate collection system. The last half of Section II must have an acceptable ground water impact as discussed above.

Conceptualization

A detailed description of the hydrogeologic site investigation will not be presented here. An accurate and appropriate site

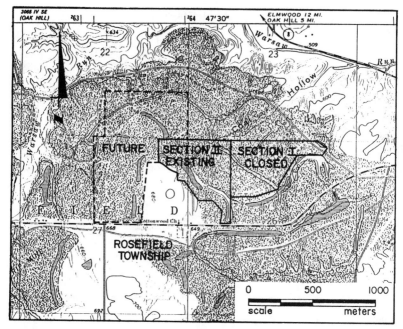

FIGURE 1.   Peoria Landfill site map.

investigation is essential because the entire modeling process is based
on it.

Information used in the hydrogeologic site investigation included
a literature review, coal mine boring logs, coal mining maps, soil
borings, monitoring wells, piezometers, as-built operational data,
laboratory soil testing, slug tests of monitoring wells, topographic
surveys, field reconnaissance, measurement of surface discharges, and
ground water quality testing. Figure 2 presents a cross-sectional view
of the resulting conceptual model.

The significant aspects of the conceptualization include the
following assumptions: 1) The fireclay (underclay) and shales beneath
No. 6 coal seam (mined or not) are relatively impermeable and will be
the lower boundary for modeling. 2) The interface between fireclay and

TECHNICAL COLLEGE OF THE LOWCOUNTRY
LEARNING RESOURCES CENTER
POST OFFICE BOX 1288
BEAUFORT, SOUTH CAROLINA 29901-1288

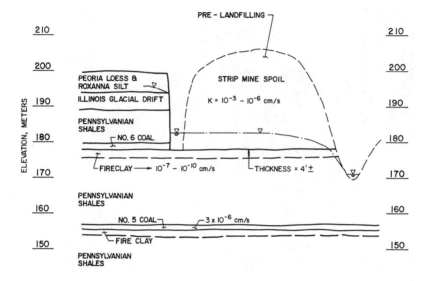

FIGURE 2. Cross-section of conceptual ground water flow model at Peoria Landfill.

mine spoil is the uppermost aquifer where horizontal ground water movements occur beneath the landfill. 3) The hydraulic properties of the mine spoil vary to such a great degree and over such small distances that porous media assumptions are the most appropriate. 4) Coal mine end cuts, coal outcrops, and surface water features serve as head-dependent boundary conditions. Lastly, it was determined that steady state flow modeling was appropriate for the objectives and conceptualization.

The objectives of the study will be achieved by dividing modeling into three components: 1) predicting leakage through the landfill liner, 2) calibrating and predicting ground water flow in the uppermost aquifer, and 3) predicting concentrations using a contaminant transport model.

Landfill Leakage

Leakage from the landfill will be predicted by the Hydrologic Evaluation of Landfill Performance (HELP) model (Shroeder et al, 1984). The model has been widely used and accepted throughout the country. It is quasi-two-dimensional; i.e., one-dimensional with variable layers. Historical precipitation and temperature data are applied and a deterministic approach is used to predict long term water balance. Precipitation is partitioned into runoff, evapotranspiration, percolation, and subsurface lateral drainage to maintain a continuous water balance.

The model accounts for synthetic liners by applying a leakage fraction (L) to the liner. This leakage factor can be correlated with the number of defects in a synthetic liner (Peyton and Schroeder, 1990). A HDPE liner with 250-1.27 cm diameter holes per hectare (6.3 m grid spacing), 0.30 m of leachate on the liner, and a 10^{-7} cm/sec clay liner would have a leakage fraction of 0.01 or 1%.

Ground Water Flow

The conceptual model indicated that a simple porous media flow model could be used, as long as the fixed head boundary conditions were addressed. Quickflow was selected on this basis (Rumbaugh, 1991). The steady state portion of the code uses "analytical elements", a relatively new advancement in flow modeling developed by Strack (1989).

The Quickflow model uses analytical functions to simulate two dimensional horizontal ground water flow, including recharge, line sinks, and sources. The head at any point in the system is calculated by using the principle of superposition and summing the affects of any number of the above functions. The major assumptions applicable for

modeling at Peoria landfill are: 1) the aquifer is isotropic and
homogeneous, 2) the aquifer is infinite in extent, 3) the reference
head in the steady state models is constant throughout all
calculations, and 4) recharge rates are constant with time.

Contaminant Transport

Horizonal GCT modeling was conducted with a two-dimensional
analytical model called PLUME2D, "Analytical Model for Transport of a
Solute Plume from Point Sources in a Uniform Two-Dimensional Ground
Water Flow Field" in the package SOLUTE written by the International
Ground Water Modeling Center (Beljin, 1989). The solution predicts
advective and dispersive transport of a nonconservative contaminant
through a semi-infinite porous media.

The ground water flow field must be modeled prior to using
PLUME2D and the seepage velocity (average linear velocity) supplied as
input. The flow field is assumed to be uniform and steady state.
Additional assumptions are: 1) uniformly porous confined aquifer, 2)
the aquifer is homogeneous, isotropic, infinite in areal extent and
constant in thickness, 3) recharge rates are negligible in relation to
uniform regional flow rate, 4) pollutants are distributed
instantaneously to the entire aquifer thickness beneath the point
sources, and 5) injection is continuous and constant.

PLUME2D will be turned on edge for this application and
constructed as a profile model. This construction is discussed below
and has implications for several of the assumptions listed above.

MODEL CONSTRUCTION

HELP

Site specific precipitation, temperature, soil characteristics, and landfill geometry data were supplied to HELP for the purposes of estimating leachate production, maximum leachate heads, and leakage (percolation) from the landfill. Figure 3 provides a representation of the model and the results. A conservative leakage factor of 1% was used for the synthetic liner. While such a leakage factor is equivalent to 250 holes per hectare, the construction quality assurance program is designed to obtain a liner with zero defects. The baseline case estimated leakage to be 0.028 cm/yr.

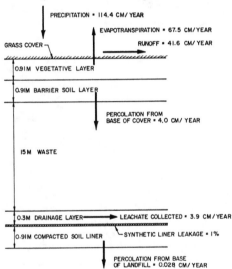

Figure 3. Representation of HELP modeling and 20 year average results.

Quickflow

The base of the model was assumed to be horizontal at elevation 178.7 m. Actual fireclay elevations varied from 178.4 m to 182.9 m, with slopes of about 1%. The chosen elevation was based on the fireclay elevation at the discharge point along Coal Hollow. The hydraulic conductivity of the mine spoil was measured to range from 1×10^{-5} cm/sec to 3×10^{-3} cm/sec. An average hydraulic conductivity of

5×10^{-4} cm/sec will be assumed beneath the entire site for the ground water flow model.

Quickflow modeling will address two cases. Case 1 is the existing condition and will be used for calibration. Case 2 is post-development and will be used to predict ground water flows after closure. The boundary conditions for each case are constant head line segments (Figure 4).

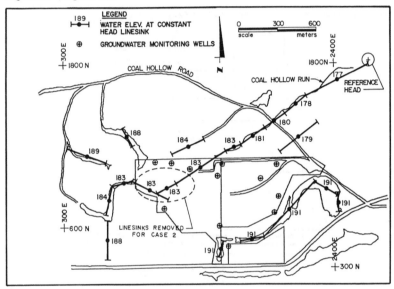

FIGURE 4. Boundary conditions for Quickflow modeling at the Peoria Landfill.

PLUME2D

The PLUME2D model was constructed as a profile model; i.e., vertical cross-section view (Figure 5). In PLUME2D nomenclature, the x-axis is oriented in the direction of ground water flow (along a streamline) and the y-axis becomes the vertical direction. The PLUME2D y-axis will be called the z-axis for the rest of this paper in recognition of it being in the vertical direction.

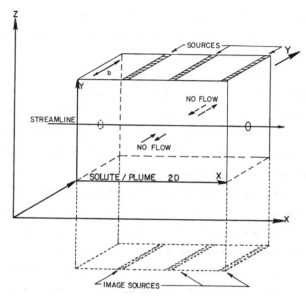

FIGURE 5. Construction of PLUME2D model at Peoria Landfill.

Since the profile model is aligned along a flowline, and PLUME2D assumes complete mixing in the 3rd-dimension, the application of a contaminant source is equivalent to an infinite line source in this 3rd dimension (y). The assumption of an infinite source perpendicular to a streamline is appropriate and conservative for a streamline located under the landfill.

It is also assumed that the streamlines are parallel to the bottom of the aquifer (fireclay). The assumption of a confined aquifer is met by application of the model in this way because orientation along a streamline results in no flow boundaries perpendicular to it (see y axis in Figure 5). The assumptions of a homogeneous and isotropic porous media are the best assumptions for the strip mine spoil. The process of mining has mixed and disoriented all the layers in the coal overburden.

Contaminant sources were applied to the surface of the water table as if holes in the landfill were routed directly to the ground water. Eight point sources spaced 30.5 m apart and placed parallel to the x-axis were used to simulate leakage through the landfill liner.

The mass rate for each source was calculated to be 0.00213 kg per year per unit width. This was based on the leakage rate calculated by HELP and assuming concentration of a leachate contaminant was 1000 mg/l. The mass rate applied to the model was doubled because flow downward into the aquifer occurs only in the -Z direction, while PLUME2D is assuming transport upwards into the landfill, which is not possible. In order to account for the impermeable boundary 6.1 m below the liner, image sources were placed 6.1 m below the impermeable base (Figure 5).

CALIBRATION

Calibration for Quickflow was based on observed heads and surface water discharges. Figure 6 presents the final calibrated model and head residuals for Case 1. Table 1 summarizes the performance of the model. Calibration targets are observed values which the model is seeking to simulate. Targets include measurement error (depth to water or survey), instrument error, monitoring well installation error, and model error involving approximations of interpolation and grid size.

Prior to modeling, the calibration targets for individual head values were determined to be observed ± 1.5 m and an average ± 0.3 m. Flux targets include measurement error, calculation error, and the possibility that the measured flow contained more than base flow. Flux targets were determined to be one order of magnitude.

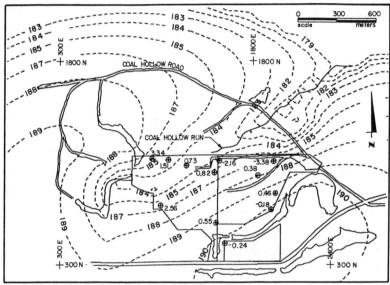

FIGURE 6.   Simulated heads for the calibrated Case 1 model at Peoria Landfill. Residuals (observed - predicted) shown for each monitoring well.

Calibration of HELP and PLUME2D was not possible. Input data for these models is based on the best information from field measurements and/or literature values. The results are carefully scrutinized for reasonableness and subject to sensitivity analysis to verify that the results are conservative and appropriate.

TABLE Performance of model against calibration targets.

Parameter	Predicted	Target Minimum	Target Observed	Target Maximum
Head Residual, m				
Minimum	0.18	0	-	1.5
Maximum	3.35	0	-	1.5
Mean	0.21	0	-	0.3
Standard Deviation	1.88	0	-	1.5
Recharge, cm/year	0.69	-	-	-
Surface Discharge, liters per minute				
Coal Hollow Run	254	42	416	4160
S301 Drain	3.75	0.38	3.79	37.9

Although several heads exceeded the target, performance was deemed acceptable after considering that the occurrences were far enough away from the area of concern to be hydraulically distant. The deviations seemed to be caused by the area having a different hydraulic conductivity or recharge than the average, or were caused by a well being close to a boundary condition not fully simulated by Quickflow. The primary cause of the latter was that Quickflow is limited in the number of constant head line segments and the head is only defined at the midpoint of the line, not all along it. The calibrated model was obtained with a minimum amount of parameter adjustment. Additional calibration on this model may yield "over fitting" of the data. Subsequent predictions (i.e., GCT) would be less accurate than those using this model.

PREDICTIONS

HELP

The results of HELP modeling are presented in Figure 3. HELP modeling (in conjunction with GCT modeling) found that a composite liner was required. Although there was a significant variation in seepage through landfill by varying the leakage fractions for the HDPE liner, the actual volume of leakage is very small. The resulting impact on the ground water is minimal even with high leakage fractions. In addition, HELP modeling demonstrated that the cap design consisting of a 0.91 m protection layer and a 0.91 m compacted liner was stable against extremes in climate.

Quickflow

Case 1 ground water flow modeling results were shown in Figure 6. Case 2 or post-closure ground water level contours are shown in Figure 7. The predicted water table is 0 to 1.83 m below the landfill except in the area of leachate collection trenches. In these areas, ground water flow will be upward through the liner where any leakage will be collected by the leachate collection system.

The streamline selected for the GCT analysis bisects the proposed area of the Section II Landfill and is also shown in Figure 7. The baseline seepage velocity was calculated based on the predicted hydraulic gradient, an average hydraulic conductivity, and estimated effective porosity:

$$V_s = \frac{Ki}{n} = 1x10^3 \frac{cm}{sec} * \frac{0.01}{0.1} * 86,400 \frac{sec}{day} = 9 \frac{cm}{day}$$

where, K = hydraulic conductivity, i = hydraulic gradient, and n = effective porosity

Due to the variability in hydraulic conductivity and uncertainty in the effective porosity, the sensitivity analysis will consider seepage velocities that vary one order of magnitude from this value.

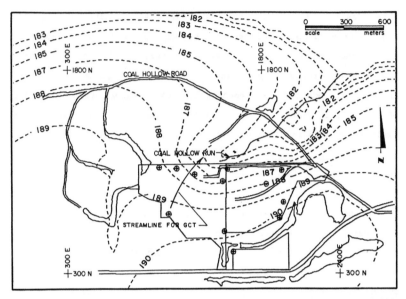

FIGURE 7. Predicted heads for Case 2, post-closure at Peoria Landfill.

The allowable error for the input variables can be calculated given this range of seepage velocities (2 orders of magnitude). First of all, the range of values for each parameter, if it is the only source of error, is as follows:

	Range
Hydraulic Conductivity, cm/s	1×10^{-4} - 1×10^{-2}
Hydraulic Gradient	0.001 - 0.10
Porosity	0.01 - 1.0

Secondly, the range can be listed if each parameter is equally in error:

	Range
Hydraulic Conductivity, cm/s	4.6×10^{-4} - 2.2×10^{-3}
Hydraulic Gradient	0.0046 - 0.22
Porosity	0.046 - 0.22

PLUME2D

The predicted concentrations for the baseline GCT simulation is shown in Figure 8. The time of prediction, 15 years, is also steady state, which occurs after approximately 11 years.

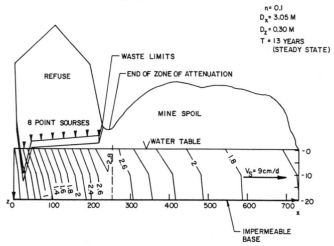

FIGURE 8. Predicted post-closure concentrations at the Peoria Landfill. Concentrations are shown as a predicted increase over background.

SENSITIVITY ANALYSIS

The results of an overall sensitivity analysis are presented in Figure 9. Although the concentrations shown are the result of PLUME2D calculations, the response of the final result to HELP and Quickflow is included. The sensitivity to ground water flow and Quickflow modeling is included in the seepage velocity input to PLUME2D. Seepage velocity decreases proportionately with hydraulic gradient and

conductivity, and inversely with effective porosity. The source
concentration term in Figure 9 is proportional to the concentration of
the contaminant in the leachate and to the rate of leakage through the
liner.

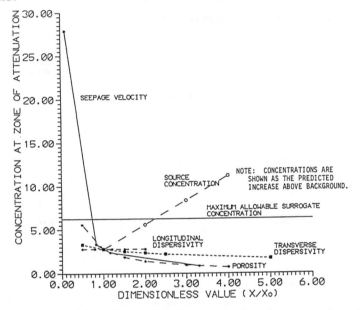

FIGURE 9. Results of sensitivity analysis for ground water
contaminant transport modeling at Peoria Landfill.

The sensitivity analysis indicated that an unacceptable ground
water impact result in two situations. The first occurs if the
contaminant leakage more than doubles from what is expected (baseline).
The second occurs when seepage velocity in the aquifer beneath the
landfill decreases more than 30% from what is expected.

DISCUSSION

The final steps of the modeling process, documentation and post-audit, will not be presented here. Documentation is contained in 4 large volumes which have been submitted to the regulatory agency. A post-audit is not possible at this stage of the project. As the landfill continues into closure and the minimum 30 years of post-closure care, post-audits are expected, especially if ground water monitoring triggers an escalation from detection monitoring to assessment monitoring. Post-audits will provide data for further calibration of the existing model or may suggest that the site requires selection of a different code and construction of a new model.

ACKNOWLEDGEMENTS

The authors acknowledge the City of Peoria and County of Peoria, who jointly own the Peoria Landfill. Their commitment to protecting the environment and public health has enabled the completion of this study.

REFERENCES

Beljin, Milovan S. 1989. SOLUTE, A Program Package of Analytical Models for Solute Transport in Ground Water. International Ground Water Modeling Center. Golden, Colorado.

Brown, David M. 1992. The fidelity fallacy. Ground Water. 30(4):4820483.

Peyton and Schroeder. May/June 1990. Evaluation of Landfill Liner Designs. Journal of Environmental Engineer. Vol. 116. No. 3.

Rumbaugh, James O. 1991. Quickflow - analytical 2D ground water flow model. Geraghty and Miller. Reston, Virginia.

Schroeder, P. R. Morgan, J. W., Walski, T. M., and Gibson, A. C. 1984. The Hydrologic Evaluation of Landfill Performance (HELP Model). U. S. Environmental Protection Agency. EPA/530-SW-84-009 and EPA/530-SW-84-010.

Strack, O. D. 1989. Ground Water Mechanics. Prentice Hall. Englewood Cliffs, New Jersey.

Selection of a Ground Water Clean Up
Option Using Monte Carlo Analysis

Andrew C. Elmore[1]
Dinshaw N. Contractor[2]

Introduction

This paper describes an application of Monte Carlo analysis coupled with decision tree analysis for a study site in Nebraska where the ground water was contaminated with trinitrotoluene (TNT). The extent of TNT contamination was characterized by a remedial investigation, and three pump and treat alternative remedial actions were developed during a feasibility study. The maximum concentration of TNT remaining in the saturated zone at the end of each alternative project lifetime was determined stochastically using a Monte Carlo model. The Monte Carlo model randomly generated initial mass concentration fields, hydraulic conductivity fields, and retardation coefficient fields. Beta probability density functions were fitted to the output ensembles of maximum concentrations. The optimal alternative was quantitatively identified using decision tree analysis.

---

[1] Project Engineer, Woodward-Clyde Federal Services, Overland Park, Kansas 66213

[2] Head, Department of Civil Engineering and Engineering Mechanics, University of Arizona, Tucson, Arizona, 85721

Deterministic modeling of pump and treat systems to clean-up contaminated ground water does not account for uncertainty associated with heterogeneous ground water systems. Uncertainty is associated with imperfect knowledge of aquifer properties including hydraulic conductivity and specific contaminant/aquifer matrix relationships represented by the retardation factor. An additional contribution to uncertainty is unknown or incomplete site history regarding the introduction of contaminants to the ground water system. Van Rooy (1986) applied Monte Carlo analysis to a case study at Zealand, Denmark where a glacial aquifer had been contaminated by landfill leachate. The analysis of the transport phenomena using a deterministic model was in good agreement with the expected values generated by the stochastic model. El-Kadi (1987) used Monte Carlo analysis to stochastically characterize simple hypothetical pump and treat ground water alternatives, the results of which indicated that many of the results obtained using the stochastic approach could not be duplicated using deterministic means. In the area of decision analysis, Schneiter, et. al., (1984) used risk-benefit analysis to select the optimal ground water alternative for a given site. Leu and Hadley (1988) discussed the use of a decision tree model to simultaneously consider toxicological, administrative, and technical factors when choosing a clean up strategy. Balagopal (1989) described a conceptual approach for using total probable risk as a criteria when applying decision tree models to the selection of hazardous waste sites.

Monte Carlo Model

The Monte Carlo simulation model consisted of a number of modules to determine the maximum concentration of TNT in the aquifer at the end of the project lifetime of each alternative. Three modules were required to generate three independent random variables: hydraulic conductivity, mass transport retardation factor, and initial mass concentration. Two additional modules provided the numerical programs for the simulation of ground water flow and contaminant mass transport simulations. The independent random variable fields were generated according to specified probability density functions and correlation structures described below. All other flow and transport parameters were assigned constant values.

Module 1: Determination of the Hydraulic Conductivity Field

The turning bands method (TBM) was used to generate the spatially auto-correlated log normal hydraulic conductivity field with an isotropic correlation function. The TBM algorithm presented in Mantoglou and Wilson (1982) was used to generate a log normal hydraulic conductivity field using the simple exponential covariance model C

$$C(r) = \sigma^2 \exp(-br) \tag{1}$$

where σ^2 is the variance of the two dimensional process, $1/b$ is the correlation length, and r is the magnitude of the two dimensional separation vector.

Module 2: Determination of Retardation Coefficient Field

A random retardation coefficient field was simulated for each Monte Carlo simulation using a two-step procedure to generate partition coefficient factors. The first step involved the simulation of a normalized partition coefficient that was held constant over the retardation coefficient field. The second step was the generation of a random fraction of organic carbon field negatively correlated to the natural log

hydraulic conductivity field.

The contaminant mass transport code used in the model accounts for solute

mobility using retardation coefficient R which is defined by

$$R = 1 + \frac{B_d K_{oc} \text{foc}}{n} \qquad (2)$$

where B_d is the bulk density of the porous media, K_{oc} is the soil-water partition

coefficient normalized with respect to fraction of organic carbon, **foc** is the fraction

of organic carbon of the porous media, and n is the porosity. The Monte Carlo

model used constant values for the bulk density and the porosity. A graph of mass

solubility C_{sol} versus K_{oc} data points presented in Spanggord et. al. (1980) was used

to derive a stochastic expression for K_{oc}. Spanggord et. al. (1980) gave the linear

least squares fit for the data as

$$\log K_{oc} = -0.27 - 0.732 \, \log C_{sol} \qquad (3)$$

Defining $Y = \ln K_{oc}$ and $X = \ln C_{sol}$, and using the definition of natural logarithm, the

equation above may be expressed in terms of the conditional mean, $E(Y \, |X = x)$

$$E(Y|x) = -0.62 - 0.732 x \qquad (4)$$

Literature review suggested that TNT solubility is relatively constant given the

anticipated range of ground water temperatures. The solubility value of
5.4 x 10^{-4} mole/L at 20°C (Spanggord et. al., 1980) was used to calculate the mean
of $\ln K_{oc}$ as 4.89. Following the technique presented in Ang and Tang (1975), the
unbiased estimate of the variance was calculated from the Spanggord et. al. (1980)
data resulting in a standard deviation of $\ln K_{oc}$ of 1.88. The chi-square goodness-of-
fit test applied to the data. The test showed that the data fit the 10% level of
significance for degree of freedom 2 as tested against the lognormal distribution. An
algorithm given by El-Kadi (1987) was used to generate a single value of K_{oc} for
each realization according to the expression

$$\ln K_{oc} = N(4.89, 1.88) \tag{5}$$

where $N(\mu, \sigma)$ represents a normal distribution with mean μ and standard deviation σ.

Garabedian (1987) showed by spectral analysis that correlation of retardation
and hydraulic conductivity produced enhanced mass spreading for reactive solutes
when compared to non-reactive solute transport. The physical basis for the
correlation is the relationship of both hydraulic conductivity and retardation to grain
size. Decreasing grain size results in increased surface area for a given mass of
aquifer material. Garabedian (1987) states that adsorption is directly proportional to
surface area so adsorption decreases with increasing grain size. Freeze and Cherry
(1979) show that hydraulic conductivity increases with increasing grain size.
Garabedian (1987) postulated that there then should be a negative correlation between

hydraulic conductivity and partition coefficient, K_p. The partition coefficient is proportionally related to foc by the expression

$$K_p = K_{oc} foc \qquad (6)$$

The Monte Carlo model calculated foc as a linear function of the natural log of hydraulic conductivity and a perturbation term. The linear function was derived so that foc was negatively correlated to natural log hydraulic conductivity. Because the natural log of hydraulic conductivity was normally distributed, the foc field was also distributed. Equation (5) was substituted into equation (2) to negatively correlate retardation to natural log hydraulic conductivity.

Module 3: Determination of Initial Mass Concentration

The initial condition necessary for the mass transport simulation is the mass concentration distribution in the aquifer at the beginning of the simulation. For the Monte Carlo model, this initial mass concentration corresponds to the mass concentration distribution in the aquifer at the beginning of the remedial alternative project life. It was assumed that the initial mass concentration was independent of the other two random variables, hydraulic conductivity and retardation factor. The Monte Carlo model was applied at a site where the spatial distribution of TNT had been characterized by analyzing ground water samples collected from monitoring wells located within and adjacent to the contaminant plume. Water levels measured in the same wells were used to estimate ground water flow directions which were considered when characterizing the TNT distribution. The methodology used to simulate the initial mass concentrations incorporates field data, best engineering judgement, and random elements to account for uncertainty. The measured mass concentrations and the mass concentrations estimated using best engineering judgement were preserved during the generation of the random initial mass concentration field by using a conditional simulation algorithm. The conditional simulation algorithm was first described by Journel (1974) and later applied by Dagbert (1981). The conditional simulation uses the turning bands method and kriging to preserve a prescribed spatial correlation structure. An unconditional simulation with a specified covariance function $C(x,y)$ is performed using TBM. Kriged estimations with the same covariance $C(x,y)$ are then superimposed on simulated values. The resulting values have covariance $C(x,y)$ and are conditioned with respect to the prescribed concentrations. The process may be represented by

$$Z_C(x) = Z_K(x) + Z_S(x) - Z_{KS}(x) \tag{7}$$

where $Z_C(x)$ is the conditional value at location x, Z_K is the value kriged using prescribed values, Z_S is the simulated (TBM) value, and Z_{KS} is the value kriged using TBM generated values at points where prescribed values exist (Dagbert, 1981). The covariance function $C(x,y)$ used in the conditional simulation is the anisotropic exponential expression

$$C(x,y) = \sigma^2 \exp[-(h_1^2 x^2 + h_2^2 y^2)] \tag{8}$$

where $1/h_1$ and $1/h_2$ are the correlation lengths in the x and y directions respectively.

The conditional simulation parameters were calibrated for the specific application using an iterative process. The area over which contamination has been characterized by field investigations was spatially discretized corresponding to the discretization that will be used by the subsequent ground water flow and transport models. Concentrations were prescribed at the appropriate grid locations using measured concentrations at monitoring wells and concentrations estimated using the results of surface and near surface soil contamination investigations (and best engineering judgement). A number of concentration fields were conditionally simulated and the mean of the ensemble was plotted and contoured. The conditional simulation variables were adjusted iteratively so that the resulting mean contours were physically reasonable. Additional concentration values were prescribed to assist in shaping the mean contours. Maximum variability in the simulation was maintained by choosing minimum correlation lengths and prescribing the fewest concentrations necessary for physically reasonable results.

Module 4: Numerical Program for Ground Water Flow

Ground water flow was simulated by the US Geological Survey MODFLOW finite difference model (McDonald and Harbaugh, 1988). Modifications were made to include MODFLOW as a module in the Monte Carlo model. The primary modification was the conversion of MODFLOW to subprogram form. Other modifications included the addition of a statement to call the subprogram which generated the hydraulic conductivity field, and modifications to the MODFLOW input subroutine to prevent excessively large files consisting of repetitive data.

Module 5: Numerical Program for Mass Transport in the Aquifer

Ground water mass transport was simulated by Papadopoulos & Associates' MT3D v1 model (Zheng, 1990). MT3D uses the method of characteristics (MOC) analysis. Earlier researchers (Van Rooy, 1986; El-Kadi 1987) have used MOC models as a part of Monte Carlo models to simulate mass transport. Modifications were made to include MT3D as a module in the Monte Carlo model. The primary modification was the conversion of MT3D to subprogram form. Other modifications included the addition of statements to call the subprograms which generated the retardation coefficient field and the initial mass concentration field. Another statement called the subprogram which analyzed the mass concentration field at the end of the simulation period.

Application to a Nebraska Contamination Site

Site Description

The Monte Carlo model was applied to a problem at a site in central Nebraska where TNT contamination in ground water had already been characterized during a remedial investigation (RI). Soil contamination, geology, and surface and subsurface hydrology data were also collected during the RI (WCC, 1989). **Figure 1** shows the location of monitoring wells where TNT was detected and the location of TNT-contaminated surface soil which was the suspected source area.

In a companion study subsequent to the RI, a Feasibility Study (FS) was conducted to evaluate potential remedial (cleanup) options. In the FS, a number of alternatives to clean up the ground water were analyzed (WCC, 1990a). Four alternatives and a baseline (no action) alternative were retained in the FS for

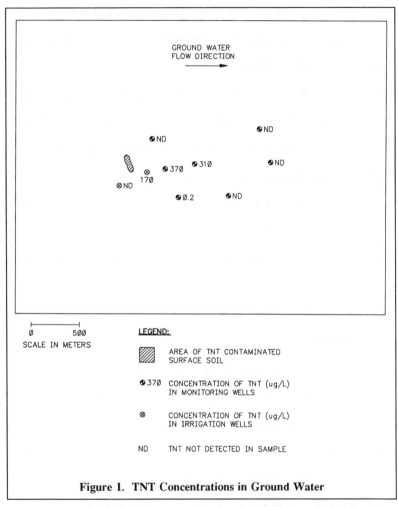

Figure 1. TNT Concentrations in Ground Water

additional consideration.   Three of the retained alternatives which pertained to pumping and treating contaminated ground water were stochastically analyzed using the Monte Carlo model.   The three alternatives are schematically illustrated in **Figure 2**.

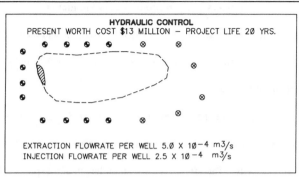

HYDRAULIC CONTROL
PRESENT WORTH COST $13 MILLION – PROJECT LIFE 20 YRS.

EXTRACTION FLOWRATE PER WELL 5.0 X 10^{-4} m3/s
INJECTION FLOWRATE PER WELL 2.5 X 10^{-4} m3/s

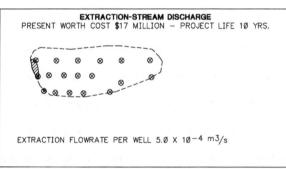

EXTRACTION-STREAM DISCHARGE
PRESENT WORTH COST $17 MILLION – PROJECT LIFE 10 YRS.

EXTRACTION FLOWRATE PER WELL 5.0 X 10^{-4} m3/s

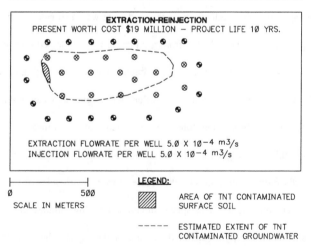

EXTRACTION-REINJECTION
PRESENT WORTH COST $19 MILLION – PROJECT LIFE 10 YRS.

EXTRACTION FLOWRATE PER WELL 5.0 X 10^{-4} m3/s
INJECTION FLOWRATE PER WELL 5.0 X 10^{-4} m3/s

Ø 500	
SCALE IN METERS	

LEGEND:

AREA OF TNT CONTAMINATED
SURFACE SOIL

----- ESTIMATED EXTENT OF TNT
CONTAMINATED GROUNDWATER

Figure 2. Remedial Alternatives

Stochastic Model Parameters Used in Monte Carlo Simulation

Input for the Monte Carlo model were derived from the RI (WCC, 1989), the FS (WCC, 1990a), a supplementary modeling study (WCC, 1990b), and other non-site related technical literature.

The hydrogeology of the site consisted of a layered glacio-fluvial aquifer. There were not enough data collected at the site to perform a statistical evaluation of hydraulic conductivity. Therefore, the correlation length and variance of the log hydraulic conductivity were estimated from the literature. Van Rooy (1986) stochastically characterized mass transport at a site where the description of the glacio-fluvial aquifer is similar to the study site. Using statistical analysis of field data, Van Rooy (1986) found the correlation length for hydraulic conductivity to be 300 m using the simple exponential model, and the variance of log hydraulic conductivity to be 0.96. Using those values as guidelines, the correlation length and variance of log hydraulic conductivity at the study site were estimated to be 244 m and 0.9 respectively. Data collected from a pumping test conducted during the RI were used to estimate the effective hydraulic conductivity at the site as 2.24×10^{-4} m/s. A study confirmed the value during ground water model flow calibration (WCC, 1990b). Therefore, the mean value of the simulated hydraulic conductivity field was assigned a value of 2.24×10^{-4} m/s. A subarea of the flow domain was discretized into a mesh with square elements with sides of 61 m (four nodes per correlation length). This spatial subdomain of flow simulation is the area where hydraulic conductivity is generated randomly and composes the entire transport domain. Elsewhere, the discretization of the flow domain was larger and the value of hydraulic conductivity was assigned a constant value of 2.24×10^{-4} m/s.

Stochastic work by Warrick and Amoozegar-Fard (1981) stated that an appropriate value for **Bd** was 1.4 g/cm3. Mackay, et. al. (1986) computed a mean value of **foc** of 0.0015 for a similar glacio-fluvial aquifer. Rao, et. al. (1986) characterized the coefficient of variation of foc as 0.20 for two sand aquifers in Florida and Georgia. The mean value of foc used in simulations of mass transport at the study site were 0.0015 and the standard deviation was 0.0003 which correlates to a coefficient of variation of 0.20. Calibration studies during deterministic ground water modeling indicated that a value of 0.20 was appropriate for effective porosity (WCC, 1990b). Substituting expected values of **Bd**, **foc**, **Koc**, and **n** into **Equation 2** yields an expected value of 9.2 for the retardation coefficient for TNT.

The RI report provides an estimate of the 100 µg/L TNT concentration in ground water as shown in **Figure 3**. The 100 µg/L estimate was used as a guide to iteratively calibrate the conditional simulation parameters. In addition to prescribing concentration values corresponding to values measured at monitoring wells, values were prescribed that do not correspond to measured concentration values to better match the estimated 100 µg/L contour. Judgement was exercised to prescribe concentration values near a suspected source area where significant TNT contamination was detected in surface and near-surface soils. It was estimated that TNT concentrations in ground water near the source were between 300 µg/L and 700 µg/L with equal probability of occurrence for all values within the range. Two points near a suspected source area were randomly assigned values for each realization

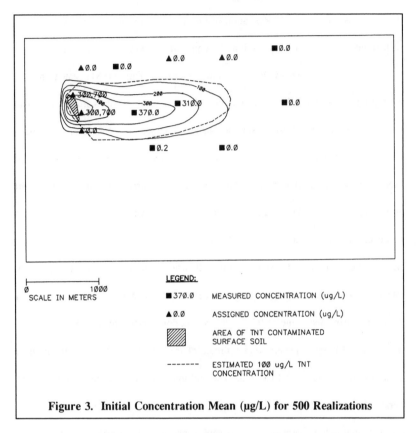

Figure 3. Initial Concentration Mean (µg/L) for 500 Realizations

according to the uniform distribution U[300,700]. The iterative approach described earlier was used to develop the following initial concentration field parameters: expected value of initial concentration equal to 4 µg/L, variance equal to 1.0, correlation length parallel to the direction of flow equal to 244 m, and correlation length normal to flow equal to 137 m. **Figure 3** shows the location of the prescribed points, the location of the area of surface soil contamination, and the mean contours for 500 realizations of initial TNT concentration.

Boundary Conditions for the Flow Model

The remaining non-random parameters necessary to simulate ground water flow and contaminant mass transport were estimated from field data and literature sources in a conventional manner with the exception of the flow boundary conditions. Natural flow boundaries such as rivers or ground water divides were not present in the vicinity of the simulated flow domain. The following procedure was implemented to generate boundary conditions for the flow simulation. A large portion of the flow domain simulated during an earlier study (WCC, 1990b) was discretized for use with a finite element flow program developed by El Didy (1986). The simulation accounted for leakage from the aquifer and seasonal recharge from irrigation. The wells for each alternative were located on the finite element grid corresponding to the well locations on the Monte Carlo model grid. The flow for each alternative was simulated for the alternative project life using the finite element model. The boundary conditions remained constant for each stress period based on heads simulated during an earlier study (WCC, 1990b) for natural conditions. The coordinates of the heads were rotated so that the coordinate system was parallel to the axes of the Monte Carlo model domain. Agricultural irrigation induced stresses on the aquifer were represented by two distinct stress periods each year. Therefore, boundary conditions for the stochastic simulation were necessary for each stress period. The total number of sets of boundary conditions is two times the number of years of the alternative project life. **Figure 4** shows the finite element simulation (prior to coordinate rotation) for the stream discharge alternative at time equal 365 days. A kriging program was used to generate a grid of head values corresponding

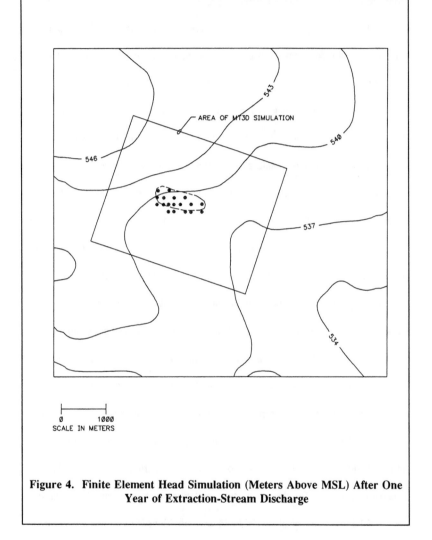

Figure 4. Finite Element Head Simulation (Meters Above MSL) After One Year of Extraction-Stream Discharge

to the Monte Carlo model grid after coordinate rotation. **Figure 5** shows the kriged head distribution using the simulated heads in **Figure 4** as input. The heads corresponding to external boundary cells were then extracted from the set of kriged

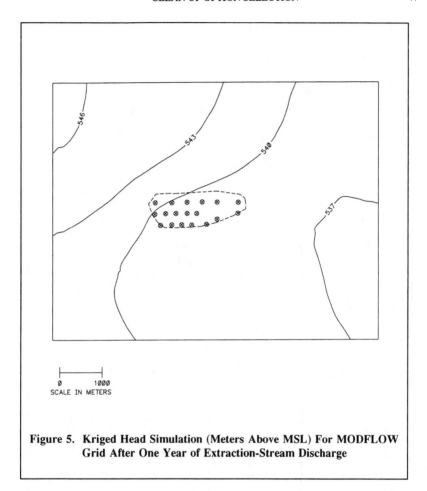

Figure 5. Kriged Head Simulation (Meters Above MSL) For MODFLOW Grid After One Year of Extraction-Stream Discharge

heads. The head values along the boundary were then adjusted to fulfill MODFLOW data input requirements. The procedure was repeated to generate general head boundary data sets for each stress period of each alternative.

Output of Monte Carlo Simulation

The Monte Carlo simulation was computationally intensive, requiring approximately nine hours of Convex 240 Superminicomputing System CPU time for every 50 realizations of the extraction-reinjection or extraction-stream discharge alternatives. Simulation of the hydraulic controls alternative took twice the CPU time because the project life was twice as long as the other alternatives' project lives. The number of realizations run for each of the alternatives is given in **Table 1**.

Table 1. Statisitics for Maximum Concentration Ensembles

	Hydraulic Controls	Extraction-Stream Discharge	Extraction-Reinjection
Number of Realizations	650	1006	1079
Mean[1] μ	327.81	235.93	272.85
Variance σ^2	29428	30033	32061
Standard Deviation σ	172	173	179
Minimum	32.5	29.5	32.7
Maximum	799.1	768.8	795.4
Mode[2]	143	65	91
Coefficient of Skewness θ	0.4590	0.8467	0.6243
[1] Maximum concentration has units of µg/L.			
[2] Estimated from histogram.			

The output from the Monte Carlo model consisted of an ensemble of maximum concentrations present in the aquifer at the end of the project life. Records of decision which describe the cleanup methods that USEPA has selected for a

Superfund site include tabulations of ground water cleanup goals which must be met for the cleanup to be complete. The cleanup goals are presented in terms of threshold contaminant concentrations which may not be exceeded in site ground water in order for the site to be "cleaned up". Beta functions were selected as being the appropriate probability density functions for the output ensembles using chi-square goodness-of-fit. The beta probability density function is given by

$$f_x(x) = \frac{1}{B(q,r)} \frac{(x-a)^{q-1}(b-x)^{r-1}}{(b-a)^{q+r-1}} \quad a \le x \le b \tag{9}$$
$$= 0 \qquad \text{elsewhere.}$$

where the beta function $B(q,r)$ is defined by

$$B(q,r) = \frac{\Gamma(q)\Gamma(r)}{\Gamma(q+r)} \tag{10}$$

The statistics for the ensembles are given in **Table 1**, and histograms and fitted Beta distributions are shown in **Figure 6**.

DECISION ANALYSIS

The Monte Carlo model results were used as input during decision analysis to select the optimal remedial alternative. The decision tree analysis which was used is a systematic framework for decision analysis under uncertainty which identifies the feasible alternatives and evaluates the respective consequences. **Figure 7** shows components of the decision tree used to evaluate the optimal remedial alternative.

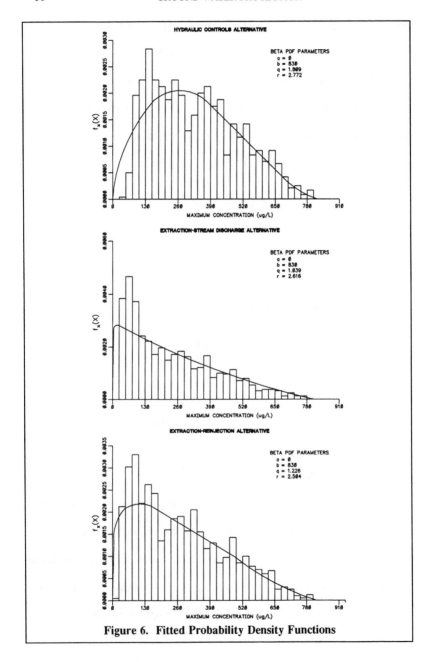

Figure 6. Fitted Probability Density Functions

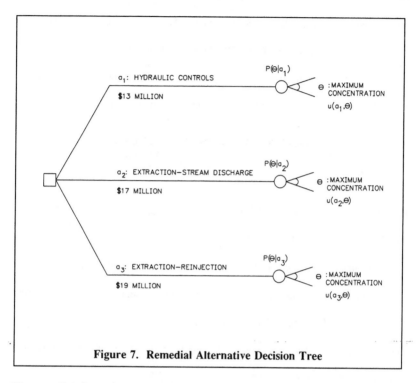

Figure 7. Remedial Alternative Decision Tree

The remedial alternatives are designated by the variable a_i; the possible outcomes θ of each alternative a_i are the continuous spectrum of maximum TNT concentrations found in the aquifer at the end of the alternative project life; and the corresponding probability assignments $P(\theta|a_i)$ are the probability density functions illustrated in **Figure 6.** The combination of monetary, human health, and regulatory consequences associated with the maximum concentration of TNT gives the utility value $u(a_i,\theta)$ corresponding to alternative a_i and outcome θ.

Regulatory consequences occur when concentrations of specific chemicals exceed U.S. Environmental Protection Agency (USEPA) or State specified values.

For this work, the regulatory concentration of TNT which was assumed to incur regulatory consequences was the USEPA Drinking Water Health Advisory level of 2 µg/L. The following expression was used to define the relationship between regulatory consequences RC and the maximum concentration of TNT x:

$$\begin{aligned} \text{Regulatory consequences} &= 0 & x \leq 2 \ \mu g/L \\ &= RC & x > 2 \ \mu g/L \end{aligned} \tag{11}$$

RC was assigned a constant value for all maximum concentrations of TNT greater than 2 µg/L for this application.

Human health consequences may be categorized as either carcinogenic or noncarcinogenic. Carcinogenic and noncarcinogenic consequences are represented by the variables CR, cancer risk, and HI, hazard index. Information presented in Life Systems, Inc. (1989) was used to develop the following relationships between CR, HI, and x:

$$CR = 1 - \exp[-x(3.1 \times 10^{-6})] \tag{12}$$

$$HI = x(0.5) \tag{13}$$

where the constants 3.1 X 10-6 and 0.5 have the units L/µg. A CR value of 6.2 X 10^{-6} is believed to be negligible in terms of a person developing cancer, and HI values less than 1.0 are believed to indicate that there is no significant risk of adverse noncarcinogenic health effects. The expected values of RC, CR, and HI for each alternative are given by:

$$E(Y) = \int_{-\infty}^{\infty} g(x) f_x(x)\, dx \tag{14}$$

where $E(Y)$ is the expected value, $g(x)$ is the expressions for RC, CR, and HI given in equations 11, 12, and 13, respectively, and $f_x(x)$ is the Beta probability density function. The numerical evaluations of $E(Y)$ are given in **Table 2**.

Table 2. Human Health and Regulatory Probabilities

	Hydraulic Controls	Extraction-Stream Discharge	Extraction-Reinjection
$E(HI)$	163.88	117.97	136.40
$E(CR)$	1.015×10^{-3}	7.3101×10^{-4}	8.4520×10^{-4}
$P(x{>}2)$	0.9999138	0.9948870	0.9982148

The expected monetary value for each alternative $E(a_i)$ was expressed as

$$E(a_1) = 13 + 163.88\,M_{HI} + 1.015 \times 10^{-3} M_{CR} + 0.999138\,M_{RC} \tag{15}$$

$$E(a_2) = 17 + 117.97\,M_{HI} + 7.301 \times 10^{-4} M_{CR} + 0.9948870\,M_{RC} \tag{16}$$

$$E(a_3) = 19 + 136.40\,M_{HI} + 8.452 \times 10^{-4} M_{CR} + 0.9982148\,M_{RC} \tag{17}$$

where $E(a_i)$ has units millions of dollars, and M_{HI}, M_{CR}, and M_{RC} are monetary values which were assigned to noncarcinogenic health consequences, carcinogenic health consequences, and regulatory consequences, respectively. The constants 13, 17, and 19 in equations 15, 16, and 17, respectively, are the capital costs of each alternative (WCC, 1990a). Using maximum expected monetary value criterion, the optimal alternative a_{opt} is the alternative with the minimum expected monetary loss:

$$d(a_{opt}) = \min_i \{E(a_i)\} \qquad (18)$$

Figure 8 shows how the optimal alternative may be found based on the values of M_{HI}, M_{CR}, and the ratio K defined as

$$K = \frac{M_{HI}}{M_{CR}} \qquad (19)$$

For $K=10^{-6}$ and values of M_{CR} below \$5,600 X 10^6, **Figure 8** shows that a_1, hydraulic controls, has the lowest monetary loss and is the optimal alternative. As the monetary value M_{CR} increases above \$5,600 X 10^6, the optimal alternative becomes a_2, extraction-stream discharge and the least desirable alternative is a_1. **Figure 8** shows that for the lower value of $K=10^{-7}$ and values of M_{CR} below \$11,600 X 10^6, the optimal alternative is a_1. As the M_{CR} increases above \$11,600 X 10^6, the optimal alternative becomes a_2. The least desirable alternative for the range of M_{CR} values plotted is always a_3, extraction-reinjection, due in part to the high capital cost of building and operating a_3.

Examination of the probabilities that each alternative would not reduce the maximum concentration of TNT in the ground water below the USEPA Drinking

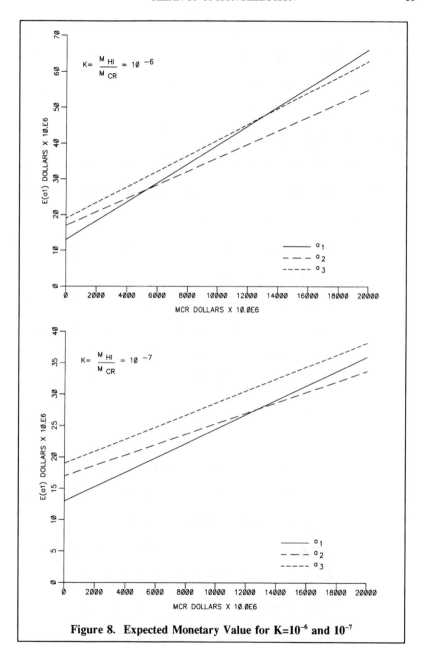

Figure 8. Expected Monetary Value for $K=10^{-6}$ and 10^{-7}

Water Health Advisory concentration of 2 µg/L (**Table 2**), indicates that the probability that any of the tree alternatives would meet the regulatory standard is not better than 0.005. The low probability of success may be improved by revising the particular design of each alternative.

Summary and Conclusions

This work demonstrates an application of Monte Carlo analysis to three pump and treat remedial alternatives using site specific data. Innovative techniques were used for the following activities during the analysis:

- Random generation of initial mass concentration fields incorporating both measured and estimated concentration values

- Random generation of retardation factor fields using random values for the partition coefficient which was sampled from an empirically derived log-normal distribution. The same distribution may be used to calculate the partition coefficient for any chemical agent which has constant solubility over the range of anticipated ground water temperatures.

- Generation of flow boundary conditions that accounted for well effects on the boundaries and allowed a spatially smaller flow domain to be analyzed to achieve computation time efficiency

- Use of the maximum concentration remaining in the aquifer at the completion of remedial activities to characterize the effectiveness of the remedial alternatives

Application of decision analysis for ground water remedial alternatives
using a decision tree model

The solution of ground water remediation problems is very complex. Technical and economic factors need to be considered and uncertainties exist in both. A considerable amount of engineering judgement is required in all aspects of simulation of the system. Despite the complexities of the prolem and the uncertainties of parameters used in the analysis, the overall methodology appears to be the most reasonable manner of addressing the problem.

REFERENCES

Ang, A.H-S., W.H. Tang. 1975. Probability concepts in engineering planning and design, volume I-basic principles. J. Wiley and Sons, New York.

Balagopal, V.G. 1989. Total probable risk analysis: a technique for quantitative risk evaluation of hazardous waste disposal options. Hazardous Waste & Hazardous Materials, 6(3), 315-325.

Dagbert, M. 1981. The simulation of space-dependent data in geology. Craig, R.G., M.L. Labovitz, eds. Future trends in geomathematics. Pion Ltd., London.

El Didy, S.M.A. 1986. Two-dimensional finite element programs for water flow and water quality in multi-aquifer systems. Ph.D. thesis. University of Arizona, Tucson.

El-Kadi, A.I. 1987. Remedial actions under variability of hydraulic conductivity. Proceedings of the NWWA conference on solving ground water problems with models. National Water Well Association, Dublin, Ohio; 705-726.

Freeze, R.A., and J.A. Cherry. 1979. Groundwater. Prentice-Hall, Inc., Englewood Cliffs, New Jersey.

Garabedian, S.P. 1987. Large-scale dispersive transport in aquifers: field experiments and reactive transport theory. Ph.D. thesis. Department of Civil Engineering, MIT.

Journel, A.G. 1974. Geostatistics for conditional simulation of ore bodies. Economic Geology, 69, 673-687.

Leu, D.J. and P.W. Hadley. 1988. The California site mitigation decision tree process: solving the "How clean should clean be?" dilemma. In Hazardous waste site management: water quality issues. National Academy Press. Washington, D.C.; 67-95.

Life Systems, Inc. 1989. Final Endangerment Assessment. Attachment to Woodward-Clyde Consultants; Final draft remedial investigation report, Hastings East Industrial Park remedial investigation/feasibility study, Hastings, Nebraska. Submitted to Department of the Army, Kansas City District, Corps of Engineers; Contract no. DACA41-87-C-0013. 28 February.

Mackay, D.M., W.P. Ball and M.G. Durant. 1986. Variability of aquifer sorption properties in a field experiment on groundwater transport of organic solutes: methods and preliminary results. Journal of Contaminant Hydrology 1; 119-132.

Mantoglou, A. and J.L. Wilson. 1982. The turning bands method for simulation of random fields using line generation by a spectral method. Water Resour. Res., 18(5). 1379-1394.

McDonald, J.M. and A.W. Harbaugh. 1988. A modular three-dimensional finite-difference ground-water flow model. Techniques of Water Resources Investigations of the U.S. Geological Survey, Book 6.

Rao, P.S.C., K.S.V. Edvardsson, L.T. Ou, R.E. Jessup, P.Nkedi-Kizza, and A.G. Hornsby. 1986. Spatial variability of pesticide sorption and degradation parameters. eds. Garner W.Y., R.C. Honeycutt and H.N. Niggs. Evaluation of pesticides in ground water, ACS symposium series no. 315. 100-115.

Schneiter, R.W., J. Dragun, and T.G. Erler. 1984. Groundwater contamination-part 3: remedial action. Chemical Engineering, 91(24); 73-78.

Spanggord, R.J., T. Mill, T-W. Chou, W.R. Mabey, J.H. Smith, and S. Lee. 1980. Environmental fate studies on certain munitions wastewater constituents. Final report Phase II - Laboratory studies. U.S. Army Medical Research and Development Command; Frederick, MD, contract no. DAMD-17-78-C-8081, September.

Van Rooy, D. 1986. Stochastic modeling of a contaminated aquifer: the unconditional approach. Nordic Hydrology, 17(4/5); 315-324.

Warrick, A.W. and A. Amoozegar-Fard. 1981. Areal predictions of water and solute flux in the unsaturated zone. U.S. Environmental Protection Agency, Robert S. Kerr Environmental Research Laboratory, contract no. EPA-600/S2-81-058. June.

Woodward-Clyde Consultants (WCC). 1989. Final draft remedial investigation report, Hastings East Industrial Park remedial investigation/feasibility study, Hastings, Nebraska. Submitted to Department of the Army, Kansas City District, Corps of Engineers; Contract no. DACA41-87-C-0013. 28 February.

Woodward-Clyde Consultants (WCC) 1990a. Advance final feasibility study report, Hastings East Industrial Park remedial investigation/feasibility study, Hastings, Nebraska. Prepared by Black and Veatch. Submitted to Department of the Army, Kansas City District, Corps of Engineers; Contract no. DACA41-87-C-0013. April 1990a.

Woodward-Clyde Consultants (WCC). 1990b. Final ground water modeling report, Hastings East Industrial Park remedial investigation/feasibility study, Hastings, Nebraska. Submitted to Department of the Army, Kansas City District, Corps of Engineers; Contract no. DACA41-87-C-0013. August.

Zheng, C. 1990. MT3D, a modular three-dimensional transport model for simulation of advection, dispersion and chemical reactions of contaminants in groundwater systems. S.S. Papadopoulos & Associates, Inc., Rockville, MD. October 17.

SIMULATED EFFECTS OF HORIZONTAL ANISOTROPY ON GROUND-WATER-FLOW PATHS AND DISCHARGE TO STREAMS

Patrick Tucci[1]

Introduction

The determination of ground-water-flow paths and the source of ground water flowing to a well are important considerations in ground-water contamination and wellhead protection studies. In homogeneous, isotropic aquifers, determination of ground-water flow paths is straightforward if an accurate potentiometric map is available. Flow lines in such an aquifer are perpendicular to equipotential lines. In anisotropic aquifers, however, determination of flow paths is not straightforward because flow lines are not perpendicular to equipotential lines. Flowlines in these aquifers are skewed in the direction of the greatest hydraulic-conductivity tensor (Fetter, 1981). The degree of skewness depends on the degree of anisotropy. In anisotropic aquifers, such as the fractured-rock aquifers of the Ridge and Valley Province of the eastern United States (Fenneman, 1938), monitoring wells located on the basis of assumed isotropic conditions, or incorrectly assumed anisotropy ratios, may not provide points for detecting contaminants transported in the ground water.

In this paper, a numerical model of a hypothetical ground-water flow system is used to demonstrate the effect of anisotropy on flow

---

[1] Hydrologist, U.S. Geological Survey, Water Resources Division, Box 25046, Denver Federal Center, MS-421, Lakewood, CO 80225

paths to a pumping well and on ground-water discharge to streams.
The simulation code used was the U.S. Geological Survey modular,
finite-difference, ground-water flow model (McDonald and Harbaugh,
1988).   A particle-tracking code (Pollock, 1989) was used to
calculate and plot flow paths.

Model description

The hypothetical ground-water flow system that was simulated
with the model is typical of fractured-rock aquifers in the Ridge and
Valley physiographic province of the United States.   Ground-water
flow in this region is commonly cited as being greater parallel to
geologic strike than perpendicular to strike, due to horizontal
anisotropy in hydraulic conductivity (Moore, 1988, p. 10).   This
anisotropy results from structural controls, such as bedding plane
partings, faults, and fractures.   In the hypothetical setting for
this exercise, the rocks strike east-west and dip to the north
(Figure 1).

Several assumptions were made to simplify simulation of the
ground-water flow system: the system is two-dimensional (all flow is
horizontal), homogeneous, of constant saturated thickness, and at
steady state, and ground-water flow to streams is at low base flow.
Recharge to the ground-water system from infiltration of rainfall is
small -- about 50 mm/yr (2 in/yr); however, this rate is similar to
rates determined for two areas near Oak Ridge, Tennessee. (Tucci,
1992; Bailey and Lee, 1991).   Three streams were included in the
model: one oriented perpendicular to strike near the western
boundary of the model (Stream C), and two oriented nearly parallel to
strike (Stream A and Stream B; Figure 1).

Constant-head boundaries were used around the periphery of the
model (Figure 2) in order to simulate ground-water inflow to the
model area, and to estimate and compare the amount of inflow as
anisotropy ratios were varied.   Streams were simulated as drains, in
that they are assumed only to receive water from the ground-water

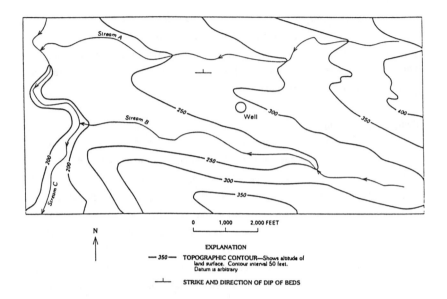

Figure 1. Physical Setting of the Problem Area.

system.   The aquifer was discretized into uniform grid blocks of 152.4 m (500 ft) on a side (Figure 2).

A transmissivity value of 92.9 m²/d (1,000 ft²/d) was assigned to the aquifer.   This value was chosen by assuming a typical hydraulic conductivity for water-bearing rocks in the Ridge and Valley of 3.0 m/d (10 ft/d), and a uniform aquifer saturated thickness of 30.5 m (100 ft). A streambed-conductance value (McDonald and Harbaugh, 1988, p. 6-4) of 92.9 m²/d (1,000 ft²/d) was uniformly assigned to each grid block simulating a stream.   This value was obtained by assuming a vertical hydraulic conductivity of the streambed of 0.03 m/d (0.1 ft/d), a streambed thickness of 0.3 m (1 ft), and an area of the stream within the grid block of 929 m² (10,000 ft²).   The values chosen for hydraulic conductivity, streambed conductance, and areal recharge are not critical to this analysis, because the scope of the analysis is generic and qualitative rather

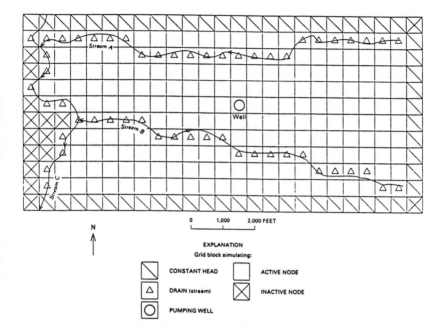

Figure 2. Model Grid and Boundaries.

than site-specific and quantitative. Use of different hydraulic parameters in the analysis would probably result in different simulated heads and water-budget components, but would not significantly affect the comparison of flow paths resulting from the use of various anisotropy ratios.

Simulated effects of anisotropy on flow paths and discharge

To demonstrate the effects of anisotropy on ground-water flow, this exercise assumes that a contaminant has been detected in a fully-penetrating well, being pumped at a rate of 6.3 L/s (100 gal/min), located near the center of the model area (Figure 2). The contaminant is conservative, and the effects of dispersion are assumed to be negligible. The source of the contaminant is unknown, and the task is to determine locations of possible sources of the contaminant. Because of the geologic structure in the study area, in which the rocks strike east-west and dip to the north, anisotropy of hydraulic conductivity of the aquifer is likely. Studies in similar hydrogeologic settings near Oak Ridge, Tennessee, indicate that hydraulic conductivity can be more than 30 times greater parallel to strike than perpendicular to strike (Lee et al., 1992, p. 591; Webster, 1976, p 15-17; Rothschild et al., 1984, p. 106; Smith and Vaughn, 1985, p. 172; Tucci, 1986, p. 5; Moore, 1989, p. 54). Model simulations were used to demonstrate ground-water-flow paths to the observation well, assuming a range of anisotropy values from 1:1 to 10:1 (strike-parallel:strike-normal), the range most commonly reported for the Oak Ridge area.

A particle-tracking program was used to calculate and plot ground-water flow paths after simulation of the ground-water system using anisotropy ratios of 1:1, 2:1, 5:1, and 10:1 (strike-parallel:strike-normal).   Twenty-five particles were distributed evenly throughout the grid block representing the observation well. These particles were tracked backward through time to their ultimate source. Particles were allowed to pass through "weak sinks", such as partially penetrating streams, as an option in the particle-tracking program (Pollock, 1989). Simulated heads were plotted for each simulation, and three flow paths were superimposed on the plot of the simulated heads. The flow paths represented the two outermost paths and the average, central flow path. Model-calculated ground-water flow from the model boundaries and to the streams was tabulated in order to examine any differences in these water-budget components due to changes in anisotropy.

Isotropic Conditions

In the simulation of isotropic conditions (anisotropy ratio = 1:1), the general direction of ground-water flow to the well is from the northeast, near the stream that is parallel to the northern boundary of the model (Figure 3). Because this stream is simulated as a drain in the ground-water system, the source of the water to the well is not the stream; however, the water originated north of the

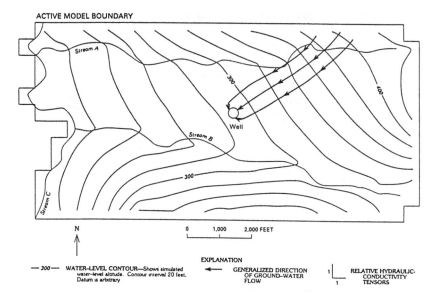

Figure 3. Simulated Water Levels and Generalized Direction of Ground-Water Flow to the Well
Under Isotropic Conditions.

stream from the constant-head boundary and passed beneath it to the
well. The calculated ground-water flow path is at right angles to
the simulated equipotential lines, as it should be for an isotropic
aquifer system.

Inflow to the ground-water system from the constant-head
boundaries is 0.28 m^3/s (10.0 ft^3/s) in this simulation, and inflow
from areal recharge is about 0.008 m^3/s (0.3 ft^3/s). This inflow is
balanced by an equal amount of outflow to the streams and pumpage
from the well (Table 1). The streams that parallel strike intercept
about 88 percent of the ground-water flow in the system, and the
stream that is oriented perpendicular to strike only intercepts about
10 percent (Table 1). The remaining 2 percent of the total inflow is
intercepted by the pumping well.

Table 1. Effects of Changing Horizontal Anisotropy Ratio on Selected
Model Results

Horizontal Anisotropy Ratio	Approximate Deviation of Flow Paths from Isotropic Conditions	Total Boundary Inflow (ft^3/s)	Percent of Flow to Strike-Parallel Streams	Percent of Flow to Strike-Normal Streams
1:1	0°	10.0	88	10
2:1	12° SE	6.5	84	14
5:1	25° SE	3.6	76	20
10:1	28° SE	2.5	67	27

Recharge = 0.3 ft^3/s
Pumpage = 0.2 ft^3/s

Anisotropic Conditions

In the simulation of anisotropic conditions, the ground-water flow paths become more skewed to the east as the anisotropy ratio is increased in the direction of strike. As the anisotropy ratio is increased, model-calculated inflow from the boundaries is reduced, the percentage of ground water intercepted by the stream oriented perpendicular to strike is increased, and simulated water levels are somewhat lowered. The source of ground water to the well is from the northeast corner of the model area when an anisotropy ratio of 2:1 is assumed (Figure 4). The average flow path deviates from that calculated for isotropic conditions by about 12 degrees southeast. Total ground-water inflow is reduced by about 35 percent from that calculated for isotropic conditions (Table 1). The percentage of total ground-water seepage to streams oriented parallel to strike is reduced slightly to 84 percent, and there is a corresponding increase (to 14 percent) in the percentage of seepage to the stream oriented perpendicular to strike (Table 1). The remaining 2 percent of the total inflow is intercepted by the pumping well. This increased percentage is due to the constant pumping rate and a decreasing boundary-inflow rate.

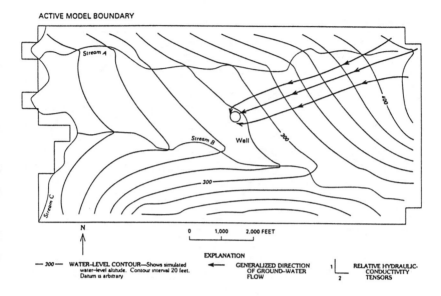

Figure 4. Simulated Water Levels and Generalized Direction of Ground-Water Flow to the Well with an Anistropy Ratio of 2:1.

When the anisotropy ratio is increased to 5:1 the ground-water flow paths are oriented in a much more easterly direction (Figure 5). The orientation of the average flow path is about 25 degrees southeast of the path calculated for isotropic conditions, and the calculated ground-water flow paths are no longer perpendicular to the simulated equipotential lines (Figure 5). The source of ground water to the well is now nearly directly east of the well, at the north-central part of the eastern model boundary. Model-calculated ground-water inflow from the boundaries is further reduced to about 0.10 m^3/s (3.6 ft^3/s; Table 1). The percentage of ground-water seepage to the streams oriented parallel to strike is reduced to 76 percent, and the percentage of seepage to the stream oriented perpendicular to strike is increased to 20 percent (Table 1). The remaining 4 percent of the total inflow is intercepted by the pumping well.

ACTIVE MODEL BOUNDARY

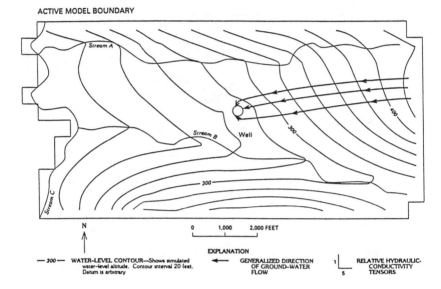

Figure 5. Simulated Water Levels and Generalized Direction of Ground-Water Flow to the Well with an Anisotropy Ratio of 5:1.

When the anisotropy ratio is increased to 10:1, the calculated ground-water flow paths to the observation well are skewed only in a slightly more easterly direction (Figure 6). The average flow path deviates from that calculated for isotropic conditions by about 28 degrees southeast. The source area of ground water to the well is similar to that for an anisotropy ratio of 5:1. Calculated ground-water inflow from the boundaries is about 0.07 m^3/s (2.5 ft^3/s), a reduction of about 73 percent from isotropic conditions (Table 1). The percentage of ground-water seepage to the streams oriented parallel to strike is further reduced to 67 percent, and the seepage to the stream oriented perpendicular to strike is increased to 27 percent (Table 1). The remaining 6 percent of the total inflow is intercepted by the pumping well.

The reduction of inflow from the constant-head boundaries, as anisotropy increases, is because most of the constant-head nodes are

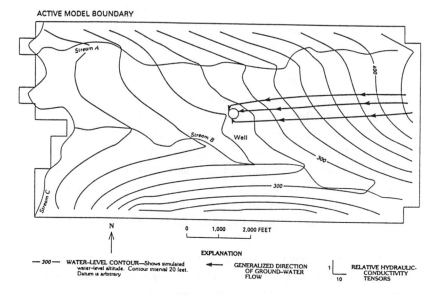

Figure 6. Simulated Water Levels and Generalized Direction of Ground-Water Flow to the Well with an Anistotropy Ratio of 10:1.

aligned parallel to strike. As anisotropy is increased, the net effect is a reduction of hydraulic conductivity perpendicular to the boundary along most of the model, thereby reducing the ground-water inflow across those boundaries.

Discussion

The effects of horizontal anisotropy of hydraulic conductivity on ground-water flow paths were readily apparent in this exercise. Simulated flow paths and source areas of ground water to the well changed dramatically as anisotropy was increased in the model. Analysis of water samples obtained from monitoring wells in the northeastern part of the model area, located on the basis of assumed isotropic conditions or an anisotropy ratio of 2:1, might not indicate contaminants moving to the well if the actual anisotropy is closer to 5:1 or more. Any model-based delineation of ground-water-flow paths, wellhead-protection areas, or placement of monitoring wells, in areas where anisotropy of hydraulic conductivity is likely should include an analysis of the potential effects of uncertainties in the orientation or magnitude of horizontal anisotropy.

As anisotropy increases, flow paths become increasingly skewed in the direction of maximum hydraulic conductivity; however, model-calculated flow paths may be insensitive to changes in anisotropy ratios beyond some threshhold value that may be site dependent. In this exercise an anisotropy ratio of 5:1 resulted in calculated ground-water-flow paths that were nearly parallel to the maximum hydraulic-conductivity tensor. Increasing the anisotropy ratio beyond 5:1 resulted in calculated flow paths that were not significantly different than those calculated with a ratio of 5:1.

Although not specifically addressed in this exercise, vertical anisotropy of hydraulic conductivity can be expected to have similar effects on ground-water-flow paths. Vertical anisotropy ratios commonly are on the order of 10:1 (horizontal to vertical direction) or greater. As the vertical anisotropy ratio increases, the ground-water-flow path should increasingly deviate from the vertical direction. A three-dimensional or cross-sectional flow model would be required to quantify the effects of vertical anisotropy on flow paths in the vertical dimension.

Model-calculated water-budget components are also affected by changes in anisotropy ratios. Calculated ground-water inflow through model boundaries was decreased as the anisotropy ratio was increased and the effective hydraulic conductivity perpendicular to the boundaries was decreased. The amount of model-calculated ground-water flow to streams that are oriented in different directions from the orientation of anisotropy was also affected by changing anisotropy ratio. A recent study of the effects of anisotropy on baseflow to streams in the Newark Basin of New Jersey also found this correlation of higher baseflow in streams oriented perpendicular to the direction of the main hydraulic-conductivity tensor (Lewis, 1992). Simulation results from that study indicated a 29 percent increase in ground-water discharge to streams oriented perpendicular to strike when simulated anisotropy increased from 1:1 to 10:1 (strike parallel:strike normal; Lewis, 1992). The significant effect of anisotropy on flow emphasizes the need for reliable measurements of water-budget components, particularly base flow to streams, for comparison to model results. Comparison of simulated heads to measured heads alone is not adequate for the unique calibration of a flow model. Comparison of measured and model-calculated budget components, in addition to heads, will provide a more comprehensive evaluation of simulation results.

Properly conducted multi-well aquifer and tracer tests can often provide valuable information on anisotropy, but such tests should be conducted on the scale of interest in order to be valid. Analysis of an aquifer or tracer test that is conducted over distances of tens or a few hundreds of feet may produce anisotropy ratios, or orientations, that may not be valid in a model used to represent an area of many square miles. The influence of a single fracture on small scale aquifer and tracer tests can be significant; however, the influence of that same fracture on the larger scale ground-water flow system may be insignificant.

References

Bailey, Z.C., and R.W. Lee, 1991. Hydrogeology and geochemistry in Bear Creek and Union Valleys near Oak Ridge, Tennessee. U.S. Geological Survey Water-Resources Investigations Report 90-4008, 72 p.

Fenneman, N.M., 1938. Physiography of eastern United States. New York, London, McGraw-Hill Book Co., Inc. pp. 195-278.

Fetter, C.W., Jr., 1981. Determination of the direction of ground-water flow. Ground Water Monitoring Review, 1 (3): 28-31.

Lee, R.R., R.H. Ketelle, J.M. Bownds, and T.A. Rizk, 1992. Aquifer analysis and modeling in a fractured, heterogeneous medium. Ground Water, 30 (4): 589-597.

Lewis, J.C., 1992. Effect of anisotropy on ground-water discharge to streams in fractured Mesozoic-basin rocks. In Hotchkiss, W.R. and A.I. Johnson, (editors). Regional Aquifer Systems of the United States, aquifers of the southern and eastern states. Bethesda, Maryland, American Water Resources Association Monograph Series 17, pp. 93-105.

McDonald, M.G., and A.W. Harbaugh, 1988. A modular three-dimensional finite-difference ground-water flow model. U.S. Geological Survey Techniques of Water-Resources Investigations, Book 6, Chapter A1, 576 p.

Moore, G.K., 1988. Concepts of groundwater occurrence and flow near Oak Ridge National Laboratory, Tennessee. Oak Ridge National Laboratory, Environmental Sciences Division Publication 3218, ORNL/TM-10969, 95 p.

Moore, G.K., 1989. Groundwater parameters and flow systems near Oak Ridge National Laboratory. Oak Ridge National Laboratory, Environmental Sciences Division Publication 3403, ORNL/TM-11368, 83 p.

Pollock, D.W., 1989. Documentation of computer programs to compute and display pathlines using results from the U.S. Geological Survey modular three-dimensional finite-difference ground-water flow model. U.S. Geological Survey Open-File Report 89-381, 188 p.

Rothschild, E.R., D.D. Huff, C.S. Haase, R.B. Clapp, B.P. Spalding, C.D. Farmer, and N.D. Farrow, 1984. Geohydrologic characterization of proposed solid waste storage area (SWSA) 7.

Oak Ridge National Laboratory, Environmental Sciences Division
Publication 2380, ORNL/TM-9314, 262 p.

Smith, E.D., and N.D. Vaughan, 1985.   Aquifer test analysis in
    nonradial flow regimes: a case study.   Ground Water, 23
    (2):167-175.

Tucci, Patrick, 1986. Ground-water flow in Melton Valley, Oak Ridge
    Reservation, Roane County, Tennessee -- preliminary model
    analysis. U.S. Geological Survey Water-Resources Investigation
    Report 85-4221, 20 p.

Tucci, Patrick, 1992.   Hydrology of Melton Valley at Oak Ridge
    National Laboratory, Tennessee. U.S. Geological Survey Water-
    Resources Investigations Report 92-4131 76 p.

Webster, D.A., 1976.   A review of hydrologic and geologic conditions
    related to the radioactive solid-waste burial grounds at Oak
    Ridge National Laboratory, Tennessee.   U.S. Geological Survey
    Open-File Report 76-727, 85p.

GROUND WATER PROTECTION IN NEW YORK STATE

Allan C. Tedrow, P.E. *

Ground Water Resources and Ground Water Use in New York

Ground water is an important resource in New York. While the major metropolitan areas in the state rely on surface water supply sources, a significant number of people in the state use ground water for water supply served through public and private systems.

There are two regions in the state of differing geologic character which is reflected in the nature of aquifers and aquifers systems: Long Island, and the remainder of the state (the "Upstate" area).

Long Island is the largest and most important aquifer system in the state. It consists of a wedge of layered deposits of unconsolidated materials up to 600 meters deep. The system has four major units which serve as sources of water supply: the Upper Glacial, the Magothy, the Jameco, and the Lloyd. The Magothy is the most important of the four.

Being completely surrounded by saline marine waters, Long Island is the sole source of drinking water for those who live on its surface in Nassau and Suffolk Counties. It was the first federally designated sole source aquifer in the state.

In the Upstate area, there is great variety in geological characteristics. Generally, the most significant ground water supplies are found in sand and gravel deposits occurring in river and stream valleys or at the sites of ancient glacial lakes. The valley deposits are often connected hydraulically to surface water streams. Wells installed in these deposits close to the surface water body can develop enhanced yields by inducing recharge from the surface waters.

Some six million people, over one-third of the population of the state, depend on ground water as a source of drinking water.

On Long Island, nearly all of the residents of Nassau and Suffolk Counties, about 2,750,000 people use ground water for drinking. In the Long Island boroughs of New York City, one-half million people are served from ground water sources. For Long Island as a whole, 90 percent of those utilizing ground water are on public systems.

In the Upstate area, three million persons rely on ground water. About one-third of them are on public systems.

Statewide, about 1.9 million cubic meters per day are supplied through public systems from ground water sources. It is estimated that another 750,000 cubic meters per day are utilized by private domestic systems. Other significant uses of ground water include industrial and commercial, 910,000 per day; and agricultural, including irrigation, 110,000. The total use of ground water is approximately 3.8 million cubic meters per day.

Early History of Ground Water Protection in New York

While a focus on quality protection of the entire ground water resource began after World War II, other efforts to address specific concerns occurred earlier. Since 1905, the DOH has had a program for adopting watershed rules and regulations on behalf of water purveyors to protect public supplies, whether surface or ground water derived. In 1930, a state law was passed establishing a regulatory program for withdrawals of ground water on Long Island. This was in response to concerns of overpumping and the salt water intrusion which resulted.

Programs to protect ground water as a resource began in

the late 1940's when the DOH, then the water pollution
control agency, began issuing permits for the construction
and operation of wastewater treatment facilities
discharging to either surface or ground water.

In the 1950's the DOH began studies to classify the
state's waters as to their highest and best use. Two
classes were developed for ground water, GA and GB. GA
indicated protection for use as a source of drinking water.
GB was intended for ground water not to be protected as a
drinking water source. Contaminated areas or places for
injection of liquid wastes apparently were to be provided
for in this second class. Numerical standards were not
adopted, but descriptive standards were. For instance,
toxics were prohibited.

In the 1960's, awareness of ground water problems
increased. Particularly on Long Island, more information on
contamination became available. Contamination of the upper
glacial aquifer by individual septic systems was found.
Metals from industrial concerns were measured in ground
water. A non-biodegradable constituent of household
detergents, ABS (alkyl benzene sulfonate) appeared in tap
water indicating the recycling of a domestic waste
component.

When ground water classifications and standards were

revised in 1967, numerical standards still were not adopted. However, numerical effluent limits for regulating permitted discharges to ground water were promulgated. As effluent limits, they were applied at the point of discharge, not in the ambient ground water. Limits were listed for 21 substances including ABS and a number of metals. Another significant change at the time of the 1967 revisions was the removal of GB as a ground water classification. This established the principle that all fresh ground waters are to be protected as sources of drinking water.

The first numerical ambient ground water standards were adopted in 1978 after the National Academy of Science published a report on drinking water. Ambient standards for some 83 substances were adopted; effluent limits for 86. For most organic substances, the numerical ambient standards equaled the effluent limits. For other substances, such as metals, the effluent limits were larger reflecting an expectation that attenuation would occur as the treated waste discharge passes through the soil.

The ambient standards and effluent limits are revised and added to as supporting information becomes available. In addition, they have been supplemented by a system of guidance values which are also used in writing discharge permits.

Development of Comprehensive Ground Water Protection Programs

Continuing concern with ground water contamination and the funding of Section 208 programs started the assembling of comprehensive ground water protection programs in the late 1970's. The work was accomplished by the state and designated regional planning agencies and resulted in two documents: the Long Island Ground Water Management Program, published June 1986; and the Upstate New York Ground Water Management Program, published May 1987.

A large number of recommendations and findings were contained in the reports. Among the most significant were the following:

1. The principle that all fresh ground water should be protected as a source of drinking water was reaffirmed. This provides that all environmental facilities should be designed and operated to prevent substances from reaching the ground water which would result in violations of ground water standard or guideline values.

2. It was recognized that because ground water quality can be lastingly impaired by intermittent or infrequent unintended discharges, there is a need for management programs beyond those used for surface water protection

to reduce contamination risks. This led to the
authorization and adoption of petroleum bulk storage and
chemical bulk storage programs.

3. There are roles in ground water protection programs
for all levels of government. While state government can
address the threat from various kinds of facilities and can
undertake remediation of problems, there is a need for
local government to be involved as well. New York has a
strong home rule tradition, and the regulation of land use
is the responsibility of municipal government. Since ground
water is affected by the nature of activities on the land
surface, a comprehensive approach to ground water
protection must include local government efforts.

4. There is a continuing need for more information on
the ground water resource. Knowledge of ground water flow
systems and better data on the location of facilities which
can affect ground water is needed as protection programs
focus increasingly on threats to water supply wells, as
is noted below.

5. While all fresh ground water is to be protected as a
source of drinking water, there are some areas for which
protection efforts should be given a heightened priority.
This concept has been called "geographic targeting." It is
a risk management approach to protection. Some ground water

systems, or parts of systems, are of greater concern based
on such factors as the presence of a water supply well, the
number of persons dependent on the ground water, the
vulnerability of the aquifer to contamination and the
yield of the aquifer. Geographic targeting can be used in
program design and in allocation of resources and efforts
for ground water protection.

Geographic Targeting

The geographic targeting concepts were further developed
in the Upstate and Long Island Ground Water Management
Programs. Specific aquifer areas were identified for
priority protection efforts.

On Long Island, various ground water flow zones were
defined during the regional Section 208 planning effort.
A deep flow recharge zone was identified as having a
particular need of protection because it serves as the
recharge area for the flow regime which is tapped by deeper
wells in Magothy Aquifer unit. Within the deep flow
recharge zone, special ground water protection areas were
also defined. These represent places with vacant lands
where opportunities still exist for affecting land use
decisions to protect high quality recharge of the deep flow
regime.

In the Upstate area, high-yielding unconsolidated deposits vulnerable to contamination were identified as being of priority concern. Eighteen "primary" aquifers were identified by the DOH as being the ones with the greatest number of people dependent on them for water supply. Of slightly lower priority are "principal" aquifers which have the same physical characteristics as primaries, but are not as intensively utilized as sources of supply at the present time.

The most critical area in the geographic targeting scheme is the water supply wellhead area. This was not given any specific physical definition in the program development effort, but was described in the Upstate report as the area of influence of a pumping well, or that portion of the recharge area where ground water moves most quickly to the well.

The geographic targeting concept has been applied by government agencies at different levels.

In a state environmental management program, it is now in regulation that new municipal landfills upstate are not to be sited in primary or principal aquifers or in wellhead protection areas. On Long Island, new or expanded landfills are not allowed in the deep flow recharge zone by law.

At the local level, county and municipal governments on Long Island control development densities according to whether the location is in the deep flow recharge area or special ground water protection areas. Also, land acquisition for ground water protection is focused on these areas.

In the cooperative data collection and information development program of the DEC and U.S. Geological Survey, mapping of primary and principal aquifers has been a priority activity for receiving allocations of funding.

The evolution of the ground water protection program has continued to develop the concept of geographic targeting through the adoption of a state wellhead protection program as required by the Safe Drinking Water Act amendments of 1986. This program, which is primarily voluntary for water purveyors, calls for design and implementation of programs to give heightened protection for the most critical part of the ground water resource - that which is to be served to the public by water supply systems.

State Ground Water Protection Program

At the present time the components of the ground water protection program implemented by state agencies include regulation, remediation, information development and outreach.

Permitting of treated wastewater discharges is a fundamental regulatory activity for ground water protection. Using ambient ground water standards and guidance values and effluent limits, the State Pollutant Discharge Elimination system writes permits for ground water discharges as well as for surface water. Monitoring of effluents and of ground water can be required by the permit.

Solid waste disposal facilities are permitted with construction standards including liners, leachate collection systems, and monitoring wells being specified. In addition, new sites are excluded from certain sensitive ground water areas as mentioned above.

The transportation, storage and disposal of hazardous wastes is regulated in compliance with federal requirements with additional stringent state standards. Ground water impacts are among the environmental effects of concern.

The storage of hazardous materials is regulated through the state petroleum bulk storage program and the chemical bulk storage program. Standards for underground tanks, leakage testing, containment and monitoring are included in the program requirements.

Pesticide sale and use is regulated by both federal and state law. With the state program, permits are required for the distribution or sale of restricted use pesticides. Permits are also needed for the purchase, possession or use of these products. All applicators of restricted use pesticides must be certified.

On-site domestic waste disposal is regulated in subdivisions through State Health Department review. The DOH also publishes standards for individual on-site systems, which are enforced through county sanitary codes and local building codes. In some places, compliance in this latter area is on a voluntary basis.

Two other state level regulatory programs are focused on the protection of water supply wells. The DOH can adopt watershed rules and regulations on behalf of water purveyors to protect the wellhead zone and the larger recharge area from activities which threaten ground water quality.

Finally, new community water supply takings are permitted by the DEC in cooperation with DOH. Conditions can be placed on the permits requiring control of lands within a set distance of the well or the adoption of watershed rules and regulations to protect ground water quality.

In summary, most of the state regulatory activity is focused on control of potential individual sources of ground water contamination. Land use controls through watershed rules and regulations are adopted only on behalf of the local water supply purveyor. Other land use controls through zoning, local ordinances, and site plan review are the province of local government.

Remediation of contamination incidents are carried out through three program routes. The first is under the state and federal superfund for inactive hazardous waste site. These provide a source of monies to investigate and initiate cleanup of sites with followup cost recovery from responsible parties. Similarly, for petroleum spills, the state oil spill fund provides a base for cleanup if the responsible party is not known or refuses initially to undertake the needed action. Violation of ground water standards provides a legal basis for enforcement to pursue remediation of a contamination incident, whatever the substance might be.

The development of information about the ground water resources of the state has been carried out through the cooperative program with the U. S. Geological Survey. Emphasis in recent years has been on the mapping of important aquifers - the primary and principal aquifers of the state.   These   have   become   more   important   as geographical targeting is utilized in protection programs at both the state and local level.

The final component of state ground water protection programs is outreach and support to local governments and water purveyors. It is in the arena of land use regulation and other wellhead protection measures that the potential for increasing effective protection is the greatest. The state has used pass-through monies available through Clean Water Act appropriations to support development of informational materials and model local ground water and wellhead protection approaches. Also, the inventorying of ground water threats as called for in the wellhead protection program is being supported as part the effort to develop local interest in ground water protection.

The future development of ground water protection will follow the lead of wellhead protection program with the fine-tuning of state programs to improve sensitivity to the recharge areas of public water supply wells. It will require improved ground water information, including portrayals of flow regimes. It will also require improved data and information systems to deal with the geography of ground water flow systems, threats of contamination and well   locations.   Finally, to enable enhanced local programs of protection, better support and information must be provided to assist those communities concerned with ground water protection to act in effective ways.

In addition, the state will be cooperating with the
U. S. Environmental Protection Agency (EPA) in developing
a Comprehensive State Ground Water Protection Program
(CSGWPP). This will lead to revisions in the state's
programs with primary emphases on better coordination of
priority-setting and operations among the various
individual programs that relate to ground water management
at the state level.

* Bureau Director (retired), Bureau of Program and
Regulatory Activities, Division of Water, New York State
Department of Environmental Conservation, 50 Wolf Road ,
Albany, NY 12233-3501

PRINCIPLES OF A GROUND WATER STRATEGY

William Whipple Jr., F.ASCE[1]
Daniel J. Van Abs, Ph.D.[2]

Updated by Daniel J. Van Abs

Introduction

This paper was prepared originally in 1989 and presented at the ASCE National Irrigation and Drainage Division Conference of that year in Newark, Delaware. Since that time, New Jersey has made substantial progress in the implementation of its *Ground Water Strategy*, though significant challenges still remain. Perhaps more importantly, the *Ground Water Strategy* remains an active policy document despite a change in administration and three changes in the organization of the lead agency responsible for its implementation. The strategy is being implemented through New Jersey's *Nonpoint Source Pollution Control Assessment and Management Program* (adopted in 1989), *Well Head Protection Program*

---

[1] In 1989, Assistant Director, Division of Water Resources, N.J. Department of Environmental Protection. Now retired from state service.

[2] In 1989, Principle Planner, Division of Water Resources, N.J. Department of Environmental Protection. Now Assistant Administrator, Office of Land and Water Planning, N.J. Department of Environmental Protection.

(Although the authors were or are employees of the New Jersey Department of Environmental Protection, the views expressed are their own and not necessarily those of the department or of the state.)

Plan (adopted in 1991), Ground Water Quality Standards (adopted in 1993), Remedial Priority System (implemented internally in 1992 and revised for public comment in 1993), State Generic Management Plan for Pesticides in Ground Water (anticipated adoption in 1994) and major new initiatives for comprehensive, watershed-based planning, permitting and management. These initiatives are discussed in the final section of this paper.

Ground water protection is a relatively new field which lacks a comprehensive federal legislative policy. Federal regulatory policies for water quality control evolved primarily for surface water, where the results of point source pollution control are reflected in relatively rapid water quality improvements. Federal programs such as the Superfund reflect this reliance on facility-oriented management. Ground water management needs are fundamentally different, and the situation urgently requires a better general approach. In the late 1980's, the states were preparing or adopting ground water strategies under EPA direction. State and federal initiatives at the time were insufficiently coordinated, and did not add up to a coherent program. Meanwhile, ubiquitous point and nonpoint pollution sources are contaminating aquifers. While facility-oriented programs and remedial programs are making some progress, much remains to be done.

New Jersey's strategy addresses these concerns. It restates or formalizes various policies, principles and procedures which are established or implicit in present practice. Significant aspects are new, such as the approach to nonpoint pollution source control programs for ground water. The most important initiatives involve coordination of water supply and hazardous waste remedial programs for contaminated wellfields. The strategy applies primarily to action by the State under its own and delegated authorities, but also has important federal implications.

Potable Water Supplies

The long term goal of water supply in New Jersey is to assure safe, adequate and affordable supplies of potable and non-potable water, at the time and place needed. This is a goal for water supply as a whole, not simply for ground water. The Department of Environmental Protection (NJDEP) regulates well drilling, water allocation, and potability of community water supplies. It plans for and sometimes funds large water supply projects, including projects to relieve stresses on major aquifers, through the *Statewide Water Supply Master Plan* and a $350 million bond issue.

The *Ground Water Strategy for New Jersey* includes several new policies not directly addressed by the existing water supply program. For instance, under State and Federal law, maximum contaminant levels for drinking water apply only to public water supplies. Under the strategy, the NJDEP will apply these same drinking water quality standards to private, residential wells, where the department is involved. This policy is of obvious importance to ground water quality restoration and the protection of private water supplies. Some municipalities and counties have already applied the Maximum Contaminant Levels (MCLs) to the testing of new private wells and property resales. The Legislature is considering statewide legislation to this end.

Also, policies are proposed to address remedies for private, residential wells. New Jersey has perhaps 250,000 private wells in use. In most cases, contaminated wells are replaced by extension of nearby water systems. However, where wells are widely scattered in rural areas, the provision of replacement supplies for contaminated wells can be extremely costly (as high as $30,000 or even more per connection). In such cases, the use of point of entry treatment systems may be endorsed as an alternative remedy. However, the department recognizes that point-of-entry systems, despite their low capital cost, are unreliable without careful monitoring and maintenance. Therefore, such systems are recommended only where more permanent and secure options (e.g., water supply lines or new wells) are infeasible or extremely costly, and where arrangements can be made to ensure maintenance and inspection by some reliable agency.

Ground Water Quality Protection

The department has adopted a "differential protection" approach, recognizing that ground water of the state varies in its value and that the resources of society are insufficient to achieve nondegradation as a statewide imperative. The strategy provides that ground water quality shall be protected by antidegradation policies, according to the natural characteristics and potential uses of the ground water, including both human and ecological uses. Absolute nondegradation is mandated only where ecological resources require it or where further degradation imperils uses.

New ground water quality standards are being promulgated with the aim of protecting ground water uses from past, present and potential discharges of pollutants.   These standards will classify ground water, establish relevant criteria for pollutants determine the appropriate antidegradation controls, and provide limited flexibility for situations where pollution mitigation is constrained by insufficient technology or extreme costs (provided that existing uses are protected from harm due to pollution).

The standards will support potable water uses in the greater part of the state, reflecting the existence of abundant ground water resources tapped by public supply systems and private wells in every region except the urban northeastern region, which relies on surface water supplies.

New Jersey is a highly urbanized state.   Though most ground water is still of excellent quality, pollution exists where people are concentrated. State policy has long been that "the polluter pays" wherever possible, and New Jersey was a leader in establishing laws, industry-funded cleanup programs and public funding to correct ground water pollution. The strategy establishes an important goal for mitigation, that of improving ground water quality to the same antidegradation levels required for new dischargers. Cleanup to the minimum quality criteria was explicitly rejected, so that ground water would not trend to a "lowest common denominator" of quality.

The strategy also breaks new ground in recognizing the value of early action to control pollution sources and contain the migration of pollution plumes. The strategy establishes a policy that source control and plume containment be effected quickly, possible using public funds in lieu of private action until such time as the responsible party agrees to act..   The

Department expects to reduce the long-term costs of pollution mitigation, and of water supply remedies, in this manner.

4.   NONPOINT SOURCE POLLUTION CONTROL

In any serious discussion of nonpoint source pollution, it is necessary to start with terminology. Legally, according to the Clean Water Act, any contamination reaching a stream by a pipe, ditch or channel is a point source of pollution. Generally speaking, small dispersed point sources and diffuse nonpoint sources may contaminate ground water as well as surface water; and all will be referred to in this paper as nonpoint sources or dispersed sources, even though, when such runoff is collected into pipes, ditch or channel, it may be classified as a point source for National Pollutant Discharge Elimination System (NPDES) control purposes. The practical significance of this approach is that is necessary to control many dispersed sources of pollution at source, prior to entry into the pipe, ditch or channel. For ground water purposes, classification of such pollution as point sources for traditional effluent discharge controls would be inappropriate.

Types of Nonpoint Source Pollution

Various nonpoint pollution sources may impair ground water quality. Agricultural practices are important in New Jersey, and even more so in some other states. Septic systems have often been included as nonpoint sources, due to the frequency of their pollution of shallow wells. Lawn care, including the use of fertilizers and pesticides, and the use of pesticides for termite control around building foundations have been identified as significant pollution sources as well. Disposal of used crankcase oil and of unused toxic and hazardous products from the home are very important. Runoff from highways and urban areas also contribute to ground water pollution.

Types of Control

Two main types of controls for nonpoint and dispersed pollution sources will be required for ground water protection. Whereas for surface sources, point source controls through NPDES permits have comprised the main regulatory approach and will continue to play a major role in the future, ground water protection must rely to a much greater extend upon best management practices (BMPs) and other controls at source (fundamentally similar to the pollution prevention approach where the goal is the minimization of pollution creation). A statewide system of BMPs must be devised, which will provide the main regulatory protection for ground water, the permit controls being relatively less important. A second type of protection for ground water, the treatment or diversion of contaminated runoff prior to infiltration or injection, will be required in limited circumstances.

The basic controlling principle is that the ground water quality standards must be used as basis for the "best management practices". However, as regards to nonpoint source pollution, it would be impracticable to establish a direct rule that no action would be allowed which would violate ground water criteria. Such a rule, if taken literally, would probably outlaw all septic tanks and many common forms of lawn care and agricultural practices. Given the inadequacy of existing information and technology, a less didactic approach will be necessary. The principles should be adopted that ground water criteria will be observed as far as practicable; and the determination as to what is practicable will be made through the various BMPs.

Area Characterization of NPS Control

Ideally from the viewpoint of pollution control, no polluting substance should be allowed anywhere. However, in view of the ubiquity of contamination from automobiles, pets, gardening, street runoff and other common aspects of civilization, priorities must be established. At least five different concepts for priority protection of areas are being advanced, more or less independently. These concepts must be dealt with in a coordinated fashion as regards nonpoint source control (as well as for protection of surface waters). The best established of these five concepts is protection of environmentally sensitive areas. Protection of trout streams, wild and scenic rivers, wildlife refuges and wetlands is a well established and familiar concept. As controls are established, they must be applied strictly in such areas. The second priority concept, which is more recent but currently very popular, is protection of coastal waters. On account of the economic and social value of bathing beaches and sea coasts, and the general fragility of marine environments, coastal areas require priority protection by best management practices or other methods.

The third concept is that of watershed protection for water supplies, which involves the necessity of avoiding unnecessary pollution by runoff or infiltration into water supply reservoirs or into streams near water supply intakes.   Quite obviously, septic tanks or gasoline service stations are contraindicated closely adjacent to water supply reservoirs.   But how far upstream should protection extend and for what facilities or practices?   The fourth concept, aquifer protection, is even less well developed.   Certain aquifer recharge areas of limited extent may provide water for wells in confined aquifers many miles away.   Because of the slow movement of ground water, the wells may not be affected for many years; however, some special precautions may be appropriate.   Again much more information is required.   The fifth concept is that of well head protection, which also involves protection of a water supply.   It has received considerable recognition because of Federal mandates for state programs in the 1986 amendments to the Safe Drinking Water Act.   Within the well head protection areas an extra degree of protection against pollution will be provided.

BMPs or permit conditions applicable statewide for various NPS and point source pollution source categories will include provisions for control of underground storage tanks, farm and garden fertilizer and pesticide use, disposition of used household chemicals, industrial septic tanks, and street and parking area runoff. For each BMP or permit control, application will be modified in accordance with areal priority. New Jersey will apply these specialized BMPs to the five areas of concern discussed above.

In summary, in the development of a ground water strategy for a state the control of nonpoint source pollution must play an important part. The extension of traditional point source permit systems, while necessary and legally very strong, must be supplemented by best management practices (BMPs) at the pollutant source if NPS pollution control is to be effective. Because of the great complexity and political sensitivity of controlling NPS pollution, a flexible approach will be necessary while the system is developed. Realistically, action extending over at least ten years will be required.

5.   HAZARDOUS WASTE AND WATER SUPPLY COORDINATION

In the water supply program, when either public or private wells are found to be contaminated in violation of maximum contaminant levels, further usage as potable water is restricted. When public funds are available, an expedited study is made to determine the extent of contamination, the local movement of ground water, the probable source of contamination and the most feasible way of providing an alternative source of water (or of treating the water to make it potable). In most cases, the probable source of contamination is not found to be one of the Federal Superfund (National Priority List) sites or "Spill Fund" hazardous waste sites handled by the State, or if it could be classified in one of these categories, the priority ranking for hazardous waste control is insufficient to initiate action. Therefore, when the water supply plan for an alternative source of water is formulated there is no way of knowing when, if ever, the source of contamination will be cleaned up or the plume contained. Also, if an attempt is made to require the originator of the pollution (the "responsible party") to pay for the new water system, their lawyers can legitimately claim that the state or the responsible party could have cleaned up the pollution at lesser cost. While all of these legal contingencies are being explored, the affected well users may be drinking bottled water for prolonged periods of time and bathing wherever they can.   That these situations are absolutely unacceptable is readily apparent but it is not easy to obtain a definitive solution within the context of state and Federal legislation and institutions.

Conceptual Approach

The essential key to the situations is that both water supply and hazardous waste programs should adopt a common priority system for dealing with situations involving proximate public health risks, that is with water supply contamination that is either already harming wells, or which can be shown to be expected at the well within the next 5 to 10 years. Investigations of such sites by both water supply and hazardous waste staffs should be initiated concurrently, as soon as it is determined that the priority is sufficiently high.   The hazardous waste study would go far enough to decide whether initial hazardous site action would consist of plume containment, early clean-up of contaminant or no action at all.   The water supply study would then be adjusted accordingly.

For example, if plume containment were to be implemented in a given case, the water supply remedial plan would need to take account only of the spread of contamination up to that time, whereas if no containment was scheduled, the new water supply system would need to provide service for additional areas to be contaminated by a further spreading of the plume. Once the study was complete, enforcement action could be taken against any "responsible party" or compensation could be requested from the State's

Spill Compensation Fund, with assurance in either case that the State's plan resulted from a comprehensive approach.

Interim Priority Rating

An initial rating of "proximate risk priority" by both hazardous waste and water supply staffs must be made to determine priority of studies on the basis of readily available information, without the more complete information obtainable only by the studies themselves. This preliminary calculation of priority points can be made by considering whether the site lies within an area of greater or less vulnerability to contamination and more or less stringent water quality goals (in accordance with ground water quality standards), and the number and capacity of wells within 2000 feet of the contamination site. The initial priority rating would be used only for ranking studies.

Final Priority Rating

After completion of both water supply and hazardous waste studies, a final priority rating will be necessary to determine the relative priority for remedial actions (if applicable) and the expenditure of public funds for such work related to proximate risks. Priority points to address proximate risks of domestic wells with contamination problems would be based on the three factors of severity of actual or predictable pollution, financial hardship, and population served.

The final priorities for publicly owned wells would also be based upon the three factors of severity, financial hardship, and population served. However, a major modification would be necessary as regards financial hardship; as the costs of remedies for polluted domestic wells are usually considerably higher than the costs associated with remedies for public water supplies.

Priorities Based on Risks Other Than
Contaminated Water in the Proximate Future

It would of course be recognized that emergency cases would be given first priority. Also, ultimate decisions as to long range clean-up of hazardous waste sites that do not involve serious proximate risks to human health would be handled differently, giving increased weight to factors such as volume and toxicity of hazardous waste at a site. Problems more complex in principle arise from proximate risks other than public health risks from contaminated drinking water. Although less frequent, such risks as fire, explosion, air pollution, radioactive radiation, physical contact with hazardous substances, and damage to fragile environments are important. Some overriding priority process could be established, whereby, after all water supply cases were established in relative priority for studies by the system described above, the NJDEP could select high risk projects of other categories to insert between the priorities of various water supply cases. Or, a more expedient solution could be developed. It could be decided that damage to fragile environments is a main focus of the water staffs, while the other named risks such as explosion and physical damage from contact from hazardous substances are of main concern to the hazardous waste staffs. Therefore, for example, if the water staff so decided, a lower ranking water supply case could be downgraded to give preference to protecting a fragile natural environment situation, and staffs could act similarly regarding situations involving a possible explosion or air contamination. In this way, special situations would be given appropriate priority, without time-wasting staff conflicts and high level involvement in the routine ranking of specific cases.

Implementation and Coordination

Most environmental programs in New Jersey are concentrated in the Department of Environment Protection. A departmental committee is recommended to assist in coordination between the diverse components. Implementation of the strategy also will require extensive coordination among other state departments that affect ground water, including Transportation and Agriculture.

In addition, the NJDEP is establishing a public advisory council to assist in the development of new initiatives in the fields of nonpoint pollution control aquifer recharge area protection and well head protection. Over time, county and municipal governments will play important roles within these same programs. Finally, the strategy emphasizes needs for ground water data acquisition and management to improve decisions, and of applied research to resolve critical technical issues.

Additional planning, analysis and regulatory development will be required for complete implementation of the Strategy. Because funds and staff are limited, the State will proceed with implementing those initiatives which provide the greatest gains in health protection, provide the greatest benefits relative to cost, or have dedicated funds from Federal or State sources.

PROGRESS FROM 1989 THROUGH 1993

The *Ground Water Strategy for New Jersey* was adopted as a Department policy document in June, 1989 during the final year of the administration of Governor Thomas Kean. It had no force in law or regulation, but established a direction for ground water program development. The Division of Water Resources was the lead agency for strategy implementation. A new administration, that of Governor Jim Florio, began the following year with an extensive change in NJDEP management and organization. Among the major changes, the Division of Water Resources was divided into functional areas (e.g., water quality permits, water supply permits, wastewater assistance, enforcement, planning, research) which were combined with similar functions of other media related programs (e.g., air, hazardous waste, coastal management) to create a department structure oriented to functional similarity rather than the affected environmental media. During the evolution of this structure, planning functions for ground water were reorganized twice. Currently, all water and land-related planning functions (including coastal zone management planning) are located within a single office (the Office of Land and Water Planning) for the first time in the NJDEP's twenty-three year history.

Despite the change in administration and reorganizations, progress has occurred in each of the major program areas described above, with some areas having more success than others. Much of the activity can be ascribed to two factors: state or federal legislation and internal program requirements that forced continued action; and a number of middle-level managers who continued work throughout the organizational changes on strategy implementation and education of new managers regarding the needs and the existing agreements as appropriate. An intra-departmental working group was established to facilitate continued progress. The public advisory committee established in 1987 for the ground water quality standards has continued work on a broad spectrum of ground water issues, including well head protection. Where success has been slow or lacking, most often the reasons center on the lack of resources, conflicts with overriding laws or regulations, or the complexity of the topics.

Domestic Potable Water Supplies

As directed by the *Ground Water Strategy*, the NJDEP began in 1989 to evaluate domestic well contamination based upon the drinking water quality standards developed for public water supply systems, wherever the NJDEP had a regulatory or non-regulatory role. Legislation passed in 1992 accepted this standard for the development of a program to test domestic wells for potability. Implementation of this new law has not begun as yet.

Ground Water Quality Standards

New ground water quality standards were adopted in early 1993 after a development process that lasted more than five years. Major issues included: the threshold for unacceptable risk of carcinogenicity (one in one million was adopted, reflecting legislative policy in New Jersey's Safe Drinking Water Act); the relationship between the standards and nonpoint source pollution controls (to be established through new regulations following the philosophy outlined above); and an innovative approach to badly polluted urban ground waters. Regarding the latter policy, the NJDEP decided to include such areas within a sub-category of the potable water classification, reflecting the past potability of such waters but also that pollution remedies are beyond current technologies. No further degradation is allowed (a clear repudiation of the "dead zone" approach) but remedies are primarily oriented to protection of receptors followed by natural attenuation.

Resource-Oriented Nonpoint Source and Point Source Pollution Control

The Department's *Nonpoint Source Pollution Assessment and Management Plan* was also adopted in 1989 and was expressly designed to dovetail with the initiatives and policies of the *Ground Water Strategy*. Two levels of nonpoint source pollution control were envisioned. First, the strategy called for control of certain pollution sources categorically and statewide, such as construction erosion and stormwater. Implementation of the initial best management practices required by Section 6217 of the Coastal Zone Act Reauthorization and Amendments is consistent with this approach. Second, the control of both nonpoint and point source pollution within areas of specific concern was addressed. This latter approach marked a resurgent interest in the field of resource-based protection, which had been a focus during the late 1970's with the preparation of *Areawide Water Quality Management Plans* (pursuant to Section 208 of the Federal

Clean Water Act) but then was eclipsed by more facility-oriented laws and programs.

1) Sensitive Area Protection -- The Ground Water Quality Standards and the Surface Water Quality Standards both specifically recognize special water areas where protection of sensitive ecosystems (e.g., the New Jersey Pinelands, trout production waters) requires more stringent regulations of pollutant discharges. These areas are included within special water classifications, and the policies are implemented through discharge control programs. However, the Department is just starting to determine how to regulate or manage nonpoint source pollution with regard to such sensitive areas.

2) Coastal Area Protection -- Most pollution control activities in the coastal area have focused on surface water impacts. However, the Sewage Infrastructure Improvement Act, adopted in 1988, provides some protection to ground water. Under the act, all sewer and stormsewer lines in four coastal counties are being mapped in detail. Stormwater outfall samples are taken in dry and wet weather conditions and investigations are conducted where pollution levels exceed thresholds. A particular focus is placed on illegal and improper connections between such systems. In addition, the Department is developing model municipal stormwater management plans and nonpoint source controls for application in the four counties (and statewide). As soils are very sandy in the region, stormwater infiltration to ground water is a common technique. Therefore, these programs will benefit ground water protection.

Section 6217 of the Coastal Zone Act Reauthorization and Amendments, in addition to basic management practices, requires that additional best management practices be applied for nonpoint source controls where necessary to protect specific coastal resources. The Department is initiating the identification of necessary requirements for compliance with this mandate, which may affect a variety of both ground and surface water pollutant sources.

3) Watershed Protection -- Numerous states, the Congress and many public interests have embraced watershed-based planning and protection efforts as a logical successor to standard, facility-oriented pollution control programs. Many of the initiatives, such as the draft of S.1114, the Clean Water Act Reauthorization, are focused on surface waters as were most of the *Areawide Water Quality Management Plans*. New Jersey, with its abundant ground water supplies, focused more than most states on ground water issues in the areawide plans and still maintains a ground water focus in its continuing planning process. In a new initiative on watershed-based management, New Jersey will be combining surface and ground water quality and quantity issues with other issues of habitat protection, implementation of the *State Development and Redevelopment Plan* and infrastructure development for water supply and wastewater management. The first pilot project began in the Whippany River watershed in northern New Jersey during the fall of 1993. Other projects partly or fully encompassing this approach are being developed.

4) Aquifer Recharge Area Protection -- The Legislature in 1988 provided funding for the Department to develop a method for mapping the aquifer recharge areas of major aquifers, preparing such maps of suitable scale for municipal use, and preparing best management practices for aquifer recharge protection (both quantity and quality). All of these products are for voluntary use by municipalities, by law. The NJDEP has published a methodology for mapping ground water infiltration that can easily be used by municipalities and counties using either hand techniques or Geographic Information Systems. The recharge area maps will be prepared during 1994 and 1995. Five management practices have been published and more are under development.

5) Well Head Protection -- The *New Jersey Well Head Protection Program Plan* was adopted by the Department and approved by U.S. EPA in December, 1991. The Legislature then appropriated $1.7 million from the 1981 Water Supply Bond Fund for program implementation. The funds are for: 1) delineation of well head protection areas (WHPAs) by the New Jersey Geological Survey for all 2,400 public community water supply wells; 2) regional demonstration projects, for which grants will be awarded in 1994; and 3) outreach and training programs. WHPA delineation regulations have been proposed for public review and comment, databases are being developed to support the delineation process, and GIS-based delineations have begun in southeastern New Jersey.

Several demonstration projects have been funded by U.S. EPA either directly or using pass-through funds for projects chosen by the NJDEP. Finally, many regulatory programs of the NJDEP and also the Pinelands Commission are planning modification to their regulations that will implement the *Well Head Protection Program Plan* for the pollutant sources and land uses they regulate. Full implementation of the plan will take another four to five years.

Priorities for Hazardous Waste Sites
Affecting Water Supplies

The NJDEP has taken three significant steps to implement the *Ground Water Strategy's* policy for consistent priorities among water supply and hazardous waste site remedies. First, a reorganization moved the water supply remedial program into the Site Remediation Program so that closer coordination was possible. Second, the NJDEP developed a Remedial Priority System that ranks sites according to their composite risk to water supplies, other water resources, air quality, sensitive ecosystems, buildings and other receptors. The site risks to each of the receptors reflect concerns of toxicity, mobility and mass, so that highly mobile and toxic substances located close to public water supply wells would receive very high scores for that concern and would tend to rank high in total site risk. Under current law, the NJDEP addresses cases according to their risk ranking, and allows those sites that rank further down to pursue remedies without intensive agency oversight.

Third, the NJDEP is developing a system of "overriding" priorities for sites where pollutants have already reached receptors at concentrations that exceed thresholds. These sites are considered "Immediate Environmental Concerns" and will be mitigated through the use of rapid-response contracts. Development of these contracts (which go beyond steps of site stabilization and waste removal) has been slow due to the restrictions of oversight and accountability laws passed by the Legislature to address concerns regarding contract management. The *Ground Water Strategy* also calls for receptors close to pollution to trigger more rapid mitigation. However, this expansion of the Immediate Environmental Concern initiative is currently beyond the staff and financial resources of the NJDEP and will be considered again at a later date.

Water Supply Planning

Not directly addressed as a *Ground Water Strategy*, but implicit in many of the Strategy's policies, is a concern for water supply issues related to ground water. The NJDEP initiated in 1989 a major revision of its 1982 *Statewide Water Supply Master Plan*. The 1982 plan addressed ground water in general terms only, due to the lack of information. Nearly $20 million have been spent in the last decade on geological and hydrogeological studies to determine the availability, quality and reliability of ground water supplies in major aquifer areas. The revised *Water Supply Plan* (to be completed late in 1994) includes a major focus on ground water supplies, the potential for contamination, areas of ground water over-use, and both capital projects and management initiatives required to protect and properly manage ground water supplies.

CONCLUSION

The New Jersey Department of Environmental Protection has continued to implement its 1989 *Ground Water Strategy* despite changes in administration and structure. Apparently, the common ground developed in 1989 has provided a solid foundation for continued effort. New Jersey also has a major advantage in that nearly all relevant programs are within one cabinet-level agency. The NJDEP is now developing a Comprehensive State Ground Water Protection Program as a pilot project in cooperation with the U.S. EPA, based upon U.S. EPA's new national strategy for ground water. This program (mostly a consolidation and integration of existing efforts) will identify new steps toward ground water protection that are not covered within the *Ground Water Strategy*. As such the comprehensive program will supersede the existing strategy in the next two years. Many years of effort still remain, but the strategy has served its initial function.

Connecticut's Wellhead Protection Program
Fred S. Banach[1]

Introduction

Ground water provides drinking water to one-third of
Connecticut's population. The state has more than 1,200
community wells and 250,000 private wells drawing about
150 million gallons per day.

[1]

Connecticut has two major types of aquifers,
unconsolidated sand and gravel deposits of glacial origin
(stratified drift aquifers) and fractured bedrock.
Stratified drift aquifers, although unevenly distributed
within the state, are the primary sources of ground water
withdrawals for public supply and large commercial and
industrial uses. Bedrock aquifers underlie the entire
state and provide substantially lower yields. They are
the principal source of water tapped by private wells to
serve individual homes.

[1]Assistant Director, Planning and Standards
Division, Water Management Bureau, Connecticut
Department of Environmental Protection, 79 Elm Street,
Hartford, CT   06106-5167

Connecticut has aggressive statewide source control programs for discharges, hazardous and solid waste, fuel and chemical storage, and other controls. We have one of the first State water quality classification systems for our ground water resources. More than 90% of Connecticut's land area is classified as GA or GAA -- ground water presumed suitable for drinking without treatment. The State's diversion program and water supply planning laws are designed to protect future sources of water supply and to ensure that water use will not exceed natural recharge or jeopardize aquatic biota.

Despite these programs, wells continue to be contaminated. The vast majority of State resources is devoted to remediation of contaminated sites. Connecticut's major aquifers are shallow, unconfined, and susceptible to contamination. Connecticut is the fourth most densely populated state. The urbanized and industrialized nature of much of Connecticut has resulted in many cases of ground water contamination. Recently, 100 wells per year are found to be polluted, mostly by solvents, petroleum products and pesticides. Major sources of contaminants include underground fuel storage, poor solvent handling practices by all users, agriculture, and solid waste disposal.

Statistics on ground water contamination in Connecticut

Records kept since 1980 show:

1. Approximately 1600 wells contaminated, both public and private, providing water to roughly 250,000

people.

2.   Agricultural pesticides (primarily EDB) account for
      about 400 wells.

3.   Solvents account for most wells affected. Solvents
      have contaminated more large public wells than other
      pollutants due to their widespread use, historic
      poor handling and mobility.

Pollution is occurring despite stringent source
controls. This situation clearly identifies the need for
a program to prevent contamination as clean up efforts
are slow, expensive, litigious, costly in terms of
personal anxiety, and not fully effective in restoring
all contamination incidents.

The key to ground water protection is preventing
nonpoint sources of ground water pollution through land
use regulation. Major public wells are mostly in gently
rolling or level sand and gravel aquifers in river
valleys: prime development areas. In order to prevent
pollution in aquifer protection areas, State and local
government must work together to control both existing
and future land uses.

This widespread recognition lead to the creation in
1988 of a legislative Task Force on Aquifer Protection.
The Task Force included representatives from business,
environmental groups, municipalities, water companies,
and state government. The recommendations of the Task
Force resulted in passage of Connecticut's Aquifer
Protection Area Act in 1989. (Aquifer Protection Task

Force; Connecticut Department of Environmental Protection, 1988; 1989)

Connecticut's Aquifer Protection Area Act

The Task Force had to consider a number of issues during its two years of deliberations. Major issues included:

1. Balancing the need for strong protection of drinking water sources and the cost of contamination against the cost of protection.

2. Fairly distributing the costs of program implementation.

3. Accurately identifying threats to ground water and devising effective ways to minimize them.

Due to its complexity, the aquifer legislation has many bugs, as Connecticut Department of Enviromental Protection (DEP) well knows and many amendments attest, but on the whole, the Act does an excellent job of dealing with the issues mentioned below:

1. The General Assembly found the cost of continued ground water contamination unacceptable and the cost of protection worthwhile, but took several measures to limit the cost of protection:

   · Accurate "Level A" mapping -- high initial cost to water companies, but may prevent costly litigation later; also will prevent costly land use regulation of areas which do not supply water to the wells;

(Connecticut   Department   of   Environmental
Protection, 1991)

·   Focused protection efforts -- aquifer protection
     areas are limited to areas of contribution and
     recharge to public wells in stratified drift
     (bedrock wells and private wells have basic
     statewide protection and possibly additional
     protection in future, once current program is
     established)

·   Exempt very small water companies from mapping
     requirements.

2. The costs of the program are shared by water
   companies, state and local government, business and
   industry (which must comply with new regulations),
   farmers, and residents of aquifer protection areas.

3. The legislation provides for improved regulation of
   existing land uses and certain future uses and
   prohibition of those land uses that pose the greatest
   risks.

Delineation. The first phase of program implementation.

Utilities (including municipal water departments) with
wells in stratified drift must perform wellhead
delineations in accordance with DEP regulations,
promulgated in 1991.

·   prepare rough (Level B) maps of the areas of
     contribution and recharge areas for any well fields

in stratified drift which serve more than 1,000 people (Level B mapping for the state is complete);

· prepare detailed (Level A) maps of the areas of contribution and recharge areas for these well fields (initiated to date for approximately 25% of wellfields, mapping deadline is 1997); and

· prepare Level B and Level A maps for proposed wells identified in approved regional Water Utility Coordinating Committee Water Supply Plans.

Protection and Management

Protection and management of delineated aquifer protection areas will be achieved through a state-local partnership, modeled in part on Connecticut's successful Inland Wetlands Program. The DEP is currently drafting land use regulations, which will form the core of the aquifer protection program. The regulations must establish:

· Best management practices for existing regulated activities located in Aquifer Protection Areas;

· Best management practices for and prohibitions of regulated activities proposed to be located in Aquifer Protection Areas;

· Procedures for exempting certain activities from these regulations if they do not pose a threat to the public wells; and

· Requirements for strategic groundwater monitoring in Aquifer Protection Areas.

Existing activities will have to conform to best management practices governing handling, storage, use and disposal of potential contaminants, chemical substitution and other pollution prevention initiatives. Many small businesses unused to complying with environmental regulations will be affected, including printers, dry cleaners, service stations, workshops and labs (in schools as well as commercial establishments), and many others. Prohibitions on new activities are likely to range from major manufacturing and waste disposal to dry cleaning, vehicle service, all underground petroleum storage except possibly for number two fuel oil, and a range of other activities.

The regulations are being developed in three phases:

· Phase one of the land use regulations, expected to be adopted in 1994, will include the prohibitions on certain new activities in Aquifer Protection Areas and will require municipalities to designate Aquifer Protection Areas and implement the prohibitions.

· Phase two (expected in 1995) will include Best Management Practices for new and existing activities within Aquifer Protection Areas.

· Phase three (expected in 1996) will complete creation of municipal Aquifer Protection programs, including regulations for local program administration, technical training for municipal officials, and strategic ground water monitoring.

The preliminary mapping indicates that there will be approximately 120 aquifer protection areas that will cover no more than 4.5% of the state's land surface. Because the affected wells tend to be in prime development areas, the population living in these areas will be substantially higher than 5% of the State's total population.

Municipal Role

First, each affected municipality must authorize an existing board or agency to act as the aquifer protection agency.   Thus far, 14% of the municipalities have authorized agencies.

Second, each affected municipality must conduct an inventory of land uses in the Level B mapped area, with special attention to activities which pose greater risks of ground water contamination. DEP sent guidelines for this inventory to the affected towns, and 35% of the inventories have been completed.

Third, each affected municipality must adopt local aquifer protection regulations for land use in municipal aquifer protection area(s). These regulations must be at least as stringent as the State land use regulations. These local regulations must be approved by the commissioner of DEP, who will require that they meet State standards and who will not approve more stringent restrictions that do not reasonably relate to ground water protection.

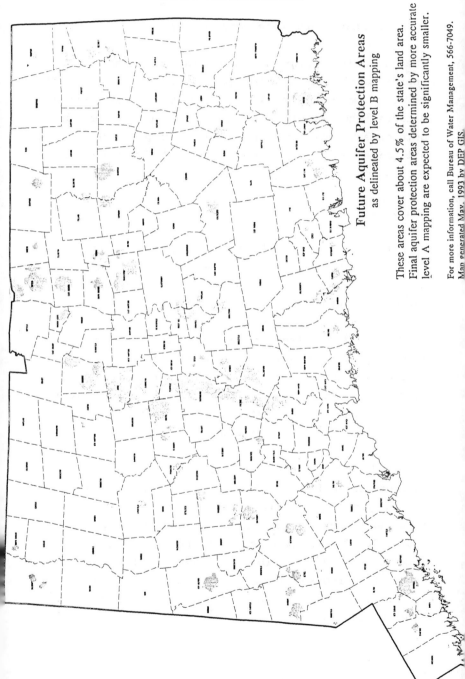

Future Aquifer Protection Areas
as delineated by level B mapping

These areas cover about 4.5% of the state's land area.
Final aquifer protection areas determined by more accurate
level A mapping are expected to be significantly smaller.

For more information, call Bureau of Water Management, 566-7049.
Map generated May, 1993 by DEP GIS.

Finally, except for facilities already regulated by individual DEP Permits, municipalities must implement the program. Although the State-local relationship in the aquifer protection program is modeled on Connecticut's inland wetlands program, aquifer protection may demand much greater staff involvement from municipalities, because it will require periodic inspection of all the regulated activities in aquifer protection areas.

If a municipality does not create or adequately implement a satisfactory aquifer protection program, the commissioner of environmental protection may revoke the local authority and administer a program for the municipality, charging the municipality for DEP's costs.

Provisions Concerning Agriculture

Farms in aquifer protection areas are to be regulated by the State rather than by the municipalities, partly because many farms cross municipal boundaries, and partly to draw on the expertise of various State and federal agriculture agencies that already work with farmers.

All farms in aquifer protection areas must submit to DEP a resources management plan laying out best management practices for groundwater protection and a schedule for implementation. The Soil and Water Conservation Districts and other agricultural organizations will provide technical assistance in developing farm resources management plans. DEP is adopting regulations for these plans, addressing manure management, pesticides storage and use, fertilizer management, and fuel storage. Adoption of this regulation is expected in 1995. DEP and SCS have also published best management practice guidelines for farmers. (Connecticut Department of Environmental Protection; Soil Conservation Service, 1993)

Other Aquifer Protection Area Program Elements

In addition to the requirement for land use regulations, the Act Concerning Aquifer Protection Areas calls on DEP to develop and implement a groundwater education program, in cooperation with the University of Connecticut Cooperative Extension Service, other State agencies, water utilities, and environmental groups; design courses to train members and staff of municipal Aquifer Protection Agencies in technical aspects of groundwater protection and implementation of the Act; use the department's geographic information system in implementation of the Act; and coordinate with other State agencies, including the departments of Transportation, Health Services, and Public Works.

Discussion

With the passage of An Act Requiring Aquifer Mapping in 1988 and An Act Concerning Aquifer Protection Areas in 1989, the State of Connecticut took a major step forward in the protection of ground water for public drinking supply. The program set up by these acts contains far reaching and very stringent controls on a wide range of land uses and places new administrative burdens on municipal government.

Clearly, getting such a program running is not an easy or quick process. However, one of the most encouraging aspects of the aquifer protection effort to date has been the nearly universal support for the protection of critical drinking water supplies, even at some cost.

Connecticut's aquifer protection program, though one of the most advanced in the country, is far from complete. Both the Aquifer Protection Task Force and the U.S. Environmental Protection Agency have stressed the need for the State to provide greater protection to the smaller public wells in bedrock in the future. Also, the continued occurrence of contamination of private wells indicates a need for better statewide ground water management. These elements of ground water protection must not be forgotten in the long and difficult process of implementing the Aquifer Protection Area Program.

Conclusion

Connecticut's Aquifer Protection Area Act provides stringent regulatory mechanisms to protect critical ground water resources surrounding wellfields.   The stringent approach is supported by rigorous hydrological investigatory methods. The Act was designed to purposely to meet the needs of Connecticut in consideration of its highly vulnerable stratified drift resources, dense developoment patterns and history of ground water contamination.   The Act will better enable DEP, munipalities, water companies and individuals better fulfill their shared environemtnal protection stewardship responsibilities to ensure high quality potable water supplies for present and future generations.

Appendix

References

1. Aquifer Protection Task Force; Connecticut Department of Environmental Protection. Report of the Aquifer Protection Task Force to the General Assembly, March 11, 1988; available from Connecticut DEP Publications, 79 Elm Street, Hartford, CT.

2. Connecticut Department of Environmental Protection. Report of the Aquifer Protection Task Force, February 15, 1989; available from Connecticut DEP Publications, 79 Elm Street, Hartford, CT.

3. Connecticut Department of Environmental Protection. Section 22a-354b-1. Regulations For Mapping Wells in Stratified Drift Aquifers To Level A Standards, April 9, 1991; available from Connecticut Department of Environmental Protection, Bureau of Water Management, 79 Elm Street, Hartford, CT.

4. Connecticut Department of Environmental Protection; Soil Conservation Service. Best Mangement Practices For Agriculture: Guidelines For Protecting Connecticut's Water Resources, 1993; available from Connecticut DEP Publications, 79 Elm Street, Hartford, CT.

APPROACHES FOR PROTECTING GROUND WATER FROM

AGRICULTURAL THREATS

Tamim M. Younos[1], A.M., ASCE, and Diana L. Weigmann[2]

INTRODUCTION

At present, more than half of the nation's population depends on ground
water for its supply of drinking water, and, in rural areas, more than 95
percent of the population relies on ground-water supplies (US EPA 1990,
EESI 1993). The quality of the nation's ground water is considered good
(USGS 1988a), but the available data, especially on the occurrence of
organic and toxic substances, generally are inadequate to determine the
full extent of ground-water contamination or define trends in ground-
water quality. However, the information base is expanding. A June
1989 report by the Office of Science and Technology Policy estimated
that federal expenditures for ground water during the five-year period

---

[1]Research Scientist, Department of Biological Systems Engineering, Virginia
Polytechnic Institute and State University, Blacksburg, Virginia, 24061.

[2]Director, Virginia Water Resources Research Center, Virginia Polytechnic
Institute and State University, Blacksburg, Virginia 24060.

through fiscal year 1990 were $890 million. More than a third of that money was used to assess the amount, quality, and hydrogeology of United States ground water. In addition, the 50 states have expanding ground-water protection programs and are adopting new approaches, such as wellhead protection.

The major sources of ground-water pollution reported to the U.S. Environmental Protection Agency (EPA) by the states are underground storage tanks, septic systems, agricultural activities, municipal landfills, surface impoundments, and abandoned hazardous waste sites (US EPA 1990). Agricultural activities are an acknowledged widespread source of pollution, and 79 percent of the states reported ground-water pollution from such practices. More than half of the states and territories identified nitrates, pesticides, volatile organic compounds, petroleum products, metals, and brine as contaminants of concern (US EPA 1990). An overview of ground-water pollution problems as related to agriculture and some corrective measures are described in a 1988 ASAE publication (White 1988).

Ground-water pollution from agricultural sources can be controlled or minimized by (1) using less chemicals, (2) employing best management practices (BMPs), (3) modifying conventional cultural practices such as implementation of integrated pest management (IPM) programs and biotechnology, (4) identifying those critical areas where ground water is more susceptible to pollution by using geographic information systems (GIS) and screening procedures, (5) conducting ground-water monitoring programs to identify and quantify pollution in affected areas, (6) implementing stringent ground-water regulations and enforcement programs, and (7) designing and adopting educational programs.

What follows is a concise discussion of the basic concepts, advantages, and limitations of these pollution control approaches.

CONVENTIONAL BMPs FOR GROUND WATER POLLUTION CONTROL

The term *best management practice* refers to a method, or combination of methods, designed for the most effective and practical control of nonpoint source (NPS) pollution. BMPs were designed and implemented originally for erosion and sediment control; protection of surface-water quality was an ancillary concern. Under the Clean Water Act and amendments by the Water Quality Act of 1987, the impacts of NPS pollution on surface-water quality are recognized. States are required to design programs using BMPs to reduce the substantial contribution of agricultural activities to the degradation of water quality. BMPs can be applied to control or minimize the pollution of ground water by agricultural chemicals. BMPs can be classified as structural measures, vegetative methods, conservation cropping, and animal-waste and nutrient management.

Structural Measures

Structural measures—terraces, grassed waterways, and diversions—have been used for many years to reduce erosion and retain topsoil. Structural BMPs are less effective for ground-water pollution control, and, in fact, structures that retain water on the surface for long periods of time, such as impoundments and level terraces, can cause an increase in infiltration volume and consequent leaching of agrichemicals into the ground water (Hanway and Laflen 1974).

Vegetative Methods

Vegetative BMPs include improving or establishing vegetative cover, critical area planting of cover crops, pasture and hayland planting, and vegetative filter strips. Planting rye on bare ground to filter out fertilizers or sediment-adsorbed pesticides is an example of a vegetative BMP. By encouraging crop nutrient uptake, cover crops may reduce the leaching of nitrate-nitrogen into the ground water (Stewart et al. 1976). Such crops may facilitate the gradual degradation of adsorbed pesticides during the winter, but the potential leaching of pesticide metabolites to ground water has not been examined under cover-crop conditions. The effectiveness of other vegetative measures, such as filter strips for removing soluble nutrients and pesticides for ground-water pollution control, is uncertain.

Conservation Cropping

The spatial and sequential arrangement of crops within an agricultural field are referred to as conservation cropping and include conservation tillage, stripcropping, and grass rotations. In conservation tillage, the most popular practice, the soil is not disturbed before planting, and ≥ 30 percent of the soil surface is covered with plant residue after planting. Herbicides are used for weed control, and negative effects of this practice on ground water have been documented. Studies have reported higher concentrations of soluble chemicals in soils under conservation tillage and the percolation of these chemicals into ground water (Baker and Johnson 1979). Increased concentrations of soluble chemicals in the infiltration water are attributed to increased moisture and nutrient leachate from plant residues acting as mulch. Mostaghimi et al. (1987) suggest that chemical losses to ground water from conservation tillage systems can be minimized by using improved application techniques. Few data are available on the impact of stripcropping, grass rotations, and other conservation cropping practices on ground-water quality.

Animal-Waste and Nutrient Management

Examples of animal-waste management systems are filter strips, no-till pasture, grazing-land protection, animal-waste storage facilities, and nutrient management. Certain of these BMPs, such as lagoons and other storage structures, offer only temporary protection because the accumulated waste from these facilities ultimately is applied to the land. The success of these methods in controlling NPS pollution depends on the design of appropriate systems for nutrient management through land treatment and application of manure.

The BMPs designed for land application of manure generally are intended only for controlling surface-water pollution. Studies (Long et al. 1975, Stewart et al. 1975) have shown that land application of manure reduces runoff and increases infiltration and the potential for ground-water pollution through improved soil structure.

Factors to consider in designing a beneficial nutrient-management program are the rate, timing, and method of manure application based on soil, topography, crop nutrient requirements, and the depth of the water table. A successful nutrient-management program includes periodic soil testing to determine fertilizer needs and manure nutrient analysis to estimate the nitrogen availability of the various manures.

BMP MODELING EFFORTS

Determining the actual sources of ground-water pollution in a mixed land-use area with different cropping and animal production systems is difficult, and much uncertainty remains on selecting the best techniques for identifying major NPS pollution-contributing areas. Also, complete agreement on how to effectively implement BMP measures has not been reached. Mathematical or computer-based NPS pollution models have proven useful as tools for pollution-load prediction and BMP implementation. Nonpoint source pollution models offer the easiest and fastest method of selecting the most practical and cost-effective BMP.

Several NPS pollution models developed during the past decade have a water-quality component (Haan et al. 1983). Certain of these models are capable of predicting the impact of agricultural activities on subsurface water or ground-water quality. One of these is the CREAMS (chemicals, runoff, and erosion from agricultural management systems) model; it is widely used and capable of simulating the ground-water impacts of BMPs (Knisel 1980). Among the user-defined management activities simulated by CREAMS are the impact of incorporating pesticides into soils, animal-waste management alternatives, and conservation tillage. The GLEAMS (ground water leaching effects on agricultural management systems) model is an extension of the CREAMS model, and can simulate leaching of pesticides and nutrients to ground water from agricultural watersheds (Leonard et al. 1987).

RUSTIC is a linked-model system developed by EPA for simulating agricultural pesticide movements in subsurface systems (Dean et al. 1989). For RUSTIC, three models, PRZM (Carsel et al. 1984), VADOFT (Huyakorn and Taylor 1987), and SAFTMOD (Huyakorn and Buckley 1988), are linked for simulating pesticide movement in the plant root zone, the intermediate vadose zone, and the saturated zone, respectively.

INTEGRATED PEST MANAGEMENT

IPM is an interdisciplinary approach to pest control, incorporating the judicious application of ecological principles, management techniques, and biological and chemical methods to maintain pest populations at tolerable levels (Bottrell 1979). IPM has been promoted in recent years because of the increased resistance of pests to synthetic pesticides, the escalating costs of synthetic pesticides, and the heightened public awareness of potential soil and water contamination and human health hazards. Among the basic tenets of IPM for reducing pesticide use are more efficient application methods, pesticide application based on economic thresholds, use of resistant crop strains, timing of field operations (planting, cultivating, harvesting), researching crop-pest ecosystems, monitoring and surveillance of pests, use of biological controls (introduction of natural enemies, preservation of predator habitats, release of sterilized male insects, use of pheromones), crop rotation, use of attractant crops, habitat diversification, use of botanicals, and destruction of pest breeding, refuge and overwintering sites (Maas et al. 1984).

Little is known about the impact of IPM on surface- and ground-water quality, and a general assumption is that the reduction in the pesticide application rate is proportional to the reduction in pesticide loss to soil and water. However, pesticide movement to ground water is strongly influenced by pesticide properties and specific field characteristics.

DATABASE DEVELOPMENT AND SCREENING TOOLS

BMPs may reduce the chemical loading to ground water, but they are effective only when climatic, geologic, and other circumstances are appropriate. Full-scale implementation of voluntary BMP programs has not yet been successful, and a more restrictive program of identifying critical areas where ground water is particularly susceptible to pollution from agricultural activities appears more feasible. Effective BMP measures can be implemented as critical areas are identified.

Identification of critical areas involves developing a database to incorporate most of the important factors related to ground-water recharge. Currently, several states have created or expanded a GIS for this purpose. By superimposing a database on both the intensity of chemical use and the GIS, critical areas are identified easily. One screening procedure developed for this purpose is DRASTIC (**D**epth to ground water, net ground water **R**echarge, **A**quifer properties, **S**oil properties, **T**opography, **I**mpact of the vandose zone, and hydraulic **C**onductivity of the aquifer).

Geographic Information Systems

Portions of agricultural watersheds, because of their land-use and geologic properties, may disproportionately affect surface- or ground-water quality. Conventional methods of site evaluation and full-scale implementation of BMPs are costly and ineffective; computerized databases such as GIS are being developed to prioritize critical areas. These critical areas contribute greater amounts of pollutants to receiving waters, and, once identified, may be targeted to introduce appropriate BMPs (e.g., Shanholtz et al. 1987).

Since the 1980s, GIS have become a valuable tool for ground-water resource management and pollution prevention and control. For example, a methodology was developed for employing GIS to study natural and anthropogenic influences on the occurrence of herbicides in ground water (USGS 1988b).

The proceedings of a 1993 international conference (Kovar and Nachtnebel 1993) included 12 articles on the application of GIS to ground-water systems, and several of these related to agricultural water-pollution control and management. In this proceedings, Battaglin (1993) discussed techniques to manage large amounts of data and high-resolution graphics to work effectively with three-dimensional tests for estimating chemical mass in ground water. Burkart and Kolpin (1993) demonstrated an application of GIS for analyzing and comparing the distribution of estimated atrazine use to that of the detection rate of atrazine in ground water.

Interfacing GIS with a hydrologic model has become a natural extension of current research activities. For example, a watershed scale evaluation system for assessing spatial variation of subsurface pesticide movement developed by Li (1993) consists of a linked-transport model component for performing simulation and a GIS component for processing spatially related data.

DRASTIC

DRASTIC is a procedure that combines seven hydraulic factors of the aquifer, as described earlier, into a numerical value that serves as an indicator of relative ground-water pollution potential for a given region of the United States (Aller et al. 1985). Each DRASTIC factor receives a rating for the geographic area under consideration. The rating then is multiplied by a weight that reflects the parameter's relative importance to pollution potential. The weighted ratings are totaled, yielding a DRASTIC score. A higher score suggests a higher degree of vulnerability to pollution.

Researchers at the U.S. Department of Agriculture (USDA) used DRASTIC to identify potential areas of ground-water contamination from pesticide and chemical fertilizer use (Nielsen and Lee 1987). However, the paucity of data available allowed only an analysis of regional trends. To make DRASTIC more useful in designing monitoring programs or taking regulatory actions, some states, including Ohio (Hallfrish and Voytek 1987) and Virginia (Wagner 1989), have developed more detailed databases for input in DRASTIC. The use of GIS and statistical approaches for more reliable data input into DRASTIC is a natural extension of these efforts.

GROUND-WATER MONITORING PROGRAMS

The hydrologic aspects of ground water have been monitored by the U.S. Geological Survey (USGS) and state agencies for many years, but less information is available about nationwide ground-water quality. Most of the states have initiated ground-water monitoring programs, and the EPA completed and published the results of a two-year nationwide survey of agrichemicals in both private and community drinking water wells throughout the United States (US EPA 1988, 1990). The National Survey of Pesticides in Drinking Water Wells (NPS) was a joint project of EPA's Office of Drinking Water and the Office of Pesticide Programs. This was the first national survey of pesticides, pesticide degradates, and nitrate in drinking water wells.

More than 1300 wells were sampled, some in each state, for 127 analytes. Of these, nitrate was most commonly detected. Based on this survey, EPA estimated that nitrate is present at or above the analytical reporting limit of 0.15 μg/L in about 52% of community wells and 57% of rural wells nationwide.

Pesticides and pesticide degradates were detected much less frequently than nitrate. Twelve of the 126 pesticides and degradates were found in the sampled wells. Based on the results of its survey, EPA estimated that 10.4% of the community wells and 4.2% of rural domestic wells in the United States contain pesticides or pesticide degradates at or above the analytical minimum reporting limit.   The two most frequently encountered pesticides were DCPA acid metabolites (degradate of dimethyl tetrachloroterphthalate) and atrazine.   In summary (listed in alphabetical order), in community wells, EPA sampled atrazine, DCPA acid metabolites, dibromochloropropane, dinoseb, hexachlorobenzene, prometon, and simazine.   In community wells, alachlor, atrazine, bentazon, DCPA acid metabolites, dibromochloropropane, ethylene dibromide, ethylene thiourea, gamma-BHC (lindane), prometon, and simazine were found.

Another source of information is the Pesticides of Ground Water Database (PGWDB), which represents a collection of ground-water monitoring studies conducted by federal, state, and local governments, the pesticide industry, and private institutions in the United States (US EPA 1992).   This database contains information from 68,824 wells, mostly drinking water wells (65,865), in 45 states.   During the 20-year period covered by the survey (1971-1991), ground water in the United States was analyzed for the presence of 302 pesticide-related compounds.   One hundred thirty-two were detected, and 35 were sampled at concentrations equal to or exceeding the MCL or lifetime health advisory.

Table 1. Summary information of PGWDB for 1971-1991.

Total number of analytes	Found in at least one well
302 pesticide-related compounds	132 pesticide-related compounds
258 parent pesticides	117 parent pesticides
45 degradates	16 degradates

Found in 100 or more wells	Found in more than 1,000 wells
23 pesticide-related compounds	7 pesticide-related compounds
21 parent pesticides	5 parent pesticides
2 degradates	2 degradates

Regulatory restrictions have been placed on 54 of the 132 pesticides detected in ground water—28 are no longer registered for use in the United States, and 27 of those with active registrations have restricted use designations. Thirty-four of the pesticides detected in ground water are, or have been, in EPA's special review process.

EPA issued an updated ground-water protection strategy in May 1991 that gives primary responsibility to the states for protecting ground water. A separate strategy issued by EPA in October 1991 described EPA's role as encouraging agricultural practices that reduce the potential for ground-water contamination, identifying pesticides that leach into ground water, using labeling or application restrictions to prevent contamination, and (if necessary) canceling federal approval of a pesticide.

LAWS AND REGULATIONS

Although EPA and USGS currently are the two main federal agencies responsible for ground-water programs, more than two-dozen agencies and offices are involved in ground-water-related activities. Sixteen federal statutes authorize programs relevant to ground-water protection (EESI 1993).

To date, Congress has passed no single law covering all aspects of ground-water management. However, several laws have provisions that directly or indirectly affect ground-water protection and management. These laws include the Clean Water Act (CWA) Amendments of 1978 and 1987 for controlling nonpoint source pollution and reducing agrichemical input to receiving waters; the Safe Drinking Water Act

(SDWA) and subsequent 1986 amendments establishing standards for certain pesticide concentrations in drinking water and programs for sole-source aquifers and wellhead protection; the Toxic Substances Control Act (TSCA) of 1976 authorizing EPA and states to restrict the use of certain pesticides in particular geographic areas to protect receiving waters from contamination; the Federal Insecticide, Fungicide, and Rodenticide Act (FIFRA) of 1978 and amendments of 1988 authorizing the EPA to enforce the pesticide registration and reregistration, reexamine pesticides as potential leachers to groundwater, and enforce the labeling requirements for proper application and storage and disposal of pesticide containers; the Resource Conservation and Recovery Act (RCRA) of 1976 regulating the disposal and treatment of hazardous materials—solvent-based, flash point of $\leq$ 60°C, aqueous, a pH of $\leq$ 2.0 or $\geq$ 12.5, and release HCN or H_2S on contact with acids.

Several bills on ground-water protection and management have been introduced in Congress but none have been signed into law as of September 1993. It is highly probable that some action will be taken by the 103rd or 104th Congress.

EDUCATIONAL PROGRAMS

A typical farmstead may include facilities such as storage tanks for chemical fertilizers, pesticides and fuels; livestock and poultry holding areas; milking centers; and silage storage silos. A septic system is generally used to dispose of household waste generated at a farmstead. At the farmstead, normal daily operations such as chemical mixing, and occasional accidents such as pesticide spills or fuel tank rupture may contribute to contamination of ground water (Hirschi et al. 1993).

A number of ground-water educational materials have been designed for individual states (for Virginia: Weigmann 1993; Weigmann and Kroehler 1988; Nickerson 1986) in the United States, and new programs are being developed each year. For example, one of these newer programs, the Farmstead Pollution Assessment System (Farm*A*Syst), is a voluntary educational and technical effort, being implemented across the nation, to protect ground-water quality and prevent contamination of household water supplies on farmsteads (Jones and Jackson 1990). The Farm*A*Syst program was originally developed and tested in Wisconsin and Minnesota through a cooperative arrangement between the EPA, USDA-Cooperative Extension Service, and USDA-Soil Conservation Service. Now it has been expanded and specifically tailored to meet the needs of those in nearly forty other states.

A typical Farm*A*Syst package contains a series of fact sheets and worksheets. Fact sheets provide background information on factors that affect farmstead water quality and legal requirements for water quality protection in farmstead environments, recommendations for corrective measures, a resource directory for additional educational and technical information, and sources for financial assistance such as state cost-share programs. Worksheets pose a series of questions framed in a risk assessment table. Questions relate to farmstead facilities and management strategies, daily operations, wellhead condition, and soils and geology of the farmstead. Based on the response to each question, a numerical ranking calculated. This ranking indicates relative water pollution risks for an individual water supply, not the actual occurrences of water contamination.

CONCLUSIONS

Ground water is a vital resource threatened by agricultural activities. Various approaches have been developed to control ground-water contamination, but few data are available to allow these approaches to be examined rigorously. Certain conventional BMPs, originally designed for sediment and erosion control, can help prevent ground-water pollution. Full-scale implementation of BMPs has not been successful nationwide; instead, computer modeling, GIS, and databases are being used to target critical areas that disproportionately affect surface- or ground-water quality. Expanded ground-water monitoring programs are providing needed data on ground-water quality, and more stringent federal laws and regulations are reducing the negative impacts of agricultural activities on water resources. Educational programs, such as Farm*A*Syst, are being designed with specific information needed to meet the particular needs for farmsteads. To date, Farm*A*Syst has been implemented in more than 80 percent of the nation.

APPENDIX. REFERENCES

Aller, L. et al. (1985). "DRASTIC: A standardized system for evaluating groundwater pollution potential using hydrologic settings." EPA/600/2-85/018, U.S. EPA, Ada, Oklahoma.

Baker, G. W., and Johnson, H.P. (1979). "The effect of tillage system on pesticide in runoff from small watersheds." TRANS Amer. Soc. Agric. Eng. 22(2), 554-558.

Battaglin, W.A. (1993). "Use of volume modelling technique to estimate agricultural chemical mass in groundwater, Minnesota, USA. In: Kovar, K., and Nachtnebel, H.P. [Eds.]. (1993). "Application of geographic information systems in hydrology and water resources management." Proceedings of an International Conference, April 1993, Vienna, Austria,

IAHS Publication No 211, IAHS Press, Institute of Hydrology, Wallingford, Oxfordshire, UK, 591-600.

Bottrell, D.R. (1979). "Integrated pest management." Council on Environmental Quality, Washington, D.C.

Burkart, M.R., and Kolpin, D.W. (1993). "Application of geographic information systems in analyzing the occurrence of atrazine in groundwater of the Mid-continental United States." In: Kovar, K., and Nachtnebel, H.P. [Eds.]. (1993). "Application of geographic information systems in hydrology and water resources management." Proceedings of an International Conference, April 1993, Vienna, Austria, IAHS Publication No 211, IAHS Press, Institute of Hydrology, Wallingford, Oxfordshire, UK, 601-610.

Carsel, R.F., et al. (1984). "User's manual for the Pesticide Root Zone Model (PRZM) -- Release I." EPA 600/3-84-109. Environmental Research Laboratory, U.S. EPA, Athens, GA, USA.

Dean, J.S., et al. (1989). "Risk of unsaturated/saturated transport and transformation of chemical concentrations (RUSTIC)." EPA 600/3-89-048b. U.S. EPA. Athens, GA, USA.

EESI (Environmental and Energy Study Institute) (1993). "1993 briefing book on environmental and energy legislation." EESI, Washington, D.C., USA.

Haan, C.T., et al. (Eds.) (1982). "Hydrologic modeling of small watersheds." Amer. Soc. Agric. Eng. Monograph No. 5., St Joseph, Michigan.

Hallfrish, M., and Voytek, J. (1987). "Groundwater pollution potential of Madison County, Ohio." Division of Water, Ohio Department of Natural Resources, Columbus, Ohio.

Hanway, J.J., and Laflen, J. M. (1974). "Nutrient losses from the outlet terraces." J. Environ. Qual., 3, 351-356.

Hirschi, M.C., et al. (1993). "50 ways farmers can protect their ground water." University of Illinois, College of Agriculture, Cooperative Extension Service C1324, Urbana-Champaign, IL, USA.

Huyakorn, P.S., and Taylor, D. (1987). "Finite element models for coupled groundwater flow and convection-dispersion." In: Gray, W.G., Pinder, G.F., and Brebbia, C.A. [Eds.]. "Finite elements in water resources." Pentech Press, London, UK, 1115-1130.

Huyakorn, P.S., and Buckley, J.E. (1988). "SAFTMOD. Saturated zone flow and transport two-dimensional finite element model." Draft report to Woodward-Clyde consultants. EPA Contract No. 68-03-6304.

Jones, S.A., and Jackson, G.W. 1990. Farmstead assessment—a strategy to prevent ground water pollution. Journal of Soil and Water Conservation, March-April, 1990, p. 236-238.

Knisel, W.G. (Ed.) (1980). "CREAMS: A field-scale model for chemicals, runoff, and erosion from agricultural management systems." U.S. Department of Agriculture, Conserv. Res. Rep. 26, Tifton, GA.

Kovar, K., and Nachtnebel, H.P. [Eds.]. (1993). "Application of geographic information systems in hydrology and water resources management." Proceedings of an International Conference, April 1993, Vienna, Austria, IAHS Publication No 211, IAHS Press, Institute of Hydrology, Wallingford, Oxfordshire, UK. 693 p.

Leonard, R.A., et al. (1987). "GLEAMS: Groundwater loading effects of agricultural management systems." TRANS Amer. Soc. Agric. Eng. 30(5), 1403-1418.

Li, W. (1993). "A subsurface water quality evaluation system for assessing NPS pollution potential by pesticides." Doctoral thesis, Virginia Polytechnic Institute and State University, Blacksburg, VA, USA. 196p.

Long, F.L., et al. (1974). "Effect of soil incorporated dairy cattle manure on runoff water quality and soil properties." J. Environ. Qual. 4(2):163-166.

Maas, R.P., et al. (1984). "Best management practices for agricultural nonpoint source control: pesticides." Natural Water Quality Evaluation Project, North Carolina State Univ., Raleigh, North Carolina.

Mostaghimi, S., et al. (1987). "Effects of tillage system, crop residue level and fertilizer application technique on losses of phosphorus and pesticides from agricultural lands." Virginia Water Resources Research Center, Bull. 157, Blacksburg, Virginia.

Nielsen, E.G., and Lee, L.K. (1987). "The magnitude and costs of groundwater contamination from agricultural chemicals: A national perspective." U.S. Department of Agriculture, Economic Research Service, Agricultural Economic Report No. 576, Washington, D.C.

Nickerson, P. 1986. Sandcastle Moats and Petunia Bed Holes. Virginia Water Resources Research Center, Virginia Polytechnic Institute and

State University, Blacksburg, Virgina, USA, 28p.

Shanholtz, V.O., et al. (1987). "Agricultural pollution potential database." Final Report to the Virginia Division of Soil and Water Conservation, Department of Agricultural Engineering, Virginia Polytechnic Institute and State Univ., Blacksburg, Virginia.

Stewart, B.A., et al. (1975). "Control of water pollutants from croplands, Vol. 1." EPA-600/2-75-026a, U.S. Environmental Protection Agency, Washington, D.C.

Stewart, B.A., et al. (1976). "Control of water pollutants from croplands, Vol. 2." EPA-600/2-75-026b, U.S. Environmental Protection Agency, Washington, D.C.

U.S. EPA (1988). "Environmental news, April 1988 issue." Office of Public Affairs, U.S. Environmental Protection Agency, Washington, D.C.

U.S. EPA (1990). "The national survey of pesticides in drinking water wells." Office of Water, Office of Pesticides and Toxic Substances, Washington, D.C., USA.

U.S. EPA (1990). "National water quality inventory: 1988 report to congress." EPA-440-4-90-003. Office of Water, Washington, D.C., USA.

U.S. EPA (1992). "Pesticides in ground water database—A compliation of monitoring studies: 1971-1991." Prevention, Pesticides, and Toxic Substances, EPA 734-12-92-001, Washington, D.C., USA.

USGS (1988a). The national water summary — 1986. Report No. 2325, Federal Center, Box 25425, Denver, CO.

USGS (1988b). "Herbicides in ground and surface water: A Mid-continent initiative." U.S. Geological Survey, Toxic Waste-Groundwater Contamination Program, Open File Report, USGS, Reston, Virginia.

Wagner, T. (1989). Virginia Water Control Board, Richmond, Virginia, Personal communication.

Weigmann, D.L. (1993). "Virginia's Groundwater." Virginia Water Resources Research Center, Virginia Polytechnic Institute and State University, Blacksburg, Virginia, USA.

Weigmann, D.L., and C.J. Kroehler. (1988). "Threats to Virginia's Groundwater." Virginia Water Resources Research Center, Virginia

Polytechnic Institute and State University, Blacksburg, Virginia, USA, 44p.

White, R.K. (1988). "Engineering approaches to solving groundwater quality problems." Public Policies Issue Report, Groundwater Quality Task Force, American Society of Agricultural Engineers, St. Joseph, MI.

Ground Water Protection - The Delaware Experience

Alan J. Farling, P.E.[1]

Introduction

Ground water in Delaware is a high quality, low cost, and readily available source of water. Approximately two-thirds of the State's population uses ground water withdrawn through public and private wells. All of the fresh water for rural use and most of the water used for irrigation and self-supplied industrial use is also derived from ground water. More than 71.9 billion liters (19 billion gallons) of ground water is withdrawn annually. Ground water also provides base flow to streams and thus is necessary for the maintenance of fish and wildlife populations and aquatic ecosystems.

Historically, ground water has been of such high quality that it has been used with little or no treatment. Since ground water is readily available in the coastal plain of Delaware, ground water development is less costly than the development of surface water sources which always require treatment.

[1]Environmental Program Administrator, Division of Water Resources, Department of Natural Resources and Environmental Control, State of Delaware, 89 Kings Highway, Dover, Delaware 19901

Additionally, wells can be placed closer to demand centers than can surface water sources thereby negating the need for extensive transmission mains. If ground water is not protected from contamination or excessive use, its abundance, high quality, low cost, and the benefits that it provides to nature will be lost for many generations to come.

It is the goal of the State of Delaware to ensure ground water of sufficient quality and quantity for the protection of public health and the preservation of significant ecological systems - now and in the future.

Delaware Environmental Protection Act

The Delaware Environmental Protection Act (DEPA) gives the Department of Natural Resources and Environmental Control (DNREC) and its Secretary broad authority to promulgate, implement, and enforce rules and regulations to protect, conserve, and assure the reasonable and beneficial uses of the State's ground water resource in the public interest. While not all inclusive, the most notable regulations issued pursuant to the DEPA address the construction of water wells, the allocation of water, the design and construction of septic systems, the siting and operation of landfills, burial of product storage tanks and pipelines, and the land application of wastes.

Comprehensive Water Resources Management Planning Program

In February 1981, then Governor Pierre S. duPont, IV, ordered DNREC to "organize, develop and implement a Comprehensive Water Resources Management Planning Program". The planning program turned out to be the most comprehensive look ever taken in Delaware at the way the State's waters were managed. A Comprehensive Water Resources Management Committee was appointed to advise the DNREC in carrying out the planning program. The Committee prepared findings and recommendations for four program areas:

>Water Management Policy
>
>Water Conservation
>
>Groundwater Quality Management
>
>Data Management.

"Water Management Policy" recommendations addressed: limits on resource development, allocation of water between competing users, consideration of prior water use, conjunctive use of ground and surface waters, water conservation requirements, a regionalization policy, and water allocation permit contents.

"Water Conservation" recommendations addressed: creating a drought index, provisions for declaration of drought warning and state of emergency due to a drought, restriction of non-essential water use during periods of declared drought emergency, protection of aquifer recharge areas, depletive water use budgeting, development of a water supply management plan, and promotion of a water conservation information and education program.

"Groundwater Quality Management" recommendations addressed: developing discharge standards so as not to degrade ground water below drinking water standards; ground water quality management being integrated with the management of water supplies; modifying septic system regulations to reduce the impact of septic system effluent on ground water; banning the sale and use of septic system cleaners which contain carcinogenic compounds; development of a program for the disposal of septage in an environmentally sound manner; expanding statewide voluntary agricultural best management activities; initiating studies to determine the best ways to use or safely dispose of poultry manure; development of criteria for the siting, operation, and abandonment of sanitary landfills; and development of criteria for the siting, operation, and abandonment of facilities handling toxic or hazardous materials or wastes.

"Data Management" recommendations addressed: manpower to input data into an electronic data management system, a well tagging system to uniquely identify a specific well, a statewide geographic information system to link water supply data with water quality data, and a semi-annual data management conference.

Regulatory Programs

There are many state and local agencies that regulate or are involved in the planning and use of ground water in Delaware. Principal agencies include the following: DNREC's Divisions of Water Resources and Air and Waste Management, Department of Agriculture, Delaware Geologic Survey, Division of Public Health, Delaware Development Office, University of Delaware Water Resources Center, New Castle County Water Resources Agency, League of Local Governments, Delaware Nature Society, Delaware Farm Bureau, Delaware Home Builders Association, and Delaware Association of Realtors.

A discussion of all of the rules and regulations governing the use and/or protection of ground water is beyond the scope of this article. However, as examples of typical regulations, discussions of the regulations governing water wells, water allocations, and septic systems follows.

The Regulations Governing the Construction of Water Wells prescribe minimum requirements for the location, design, installation, use, disinfection, modification, repair, and abandonment of all wells. A permit must be obtained for the installation of any well. The well permitting program is designed to protect the ground water by preventing pollutants from entering the ground water via improperly constructed wells. Of the 4215 well permits issued by the DNREC in 1990, 2758 were for domestic use, 88 were for public use, 34 were for irrigation use, and 798 were for monitor wells. The construction of 16% of the wells was inspected by the DNREC staff to insure compliance with well construction standards and well permit conditions. Injection wells are permitted only for heat pump recharge wells. Additionally, all well drillers are licensed by the DNREC. Obtaining a license requires a minimum of two years of experience and passing a written exam.

Water allocation permits are required for all wells withdrawing more than 189,250 liters (50,000 gallons) in any 24 hour period. The permits encourage development of regional water service facilities and interconnections between public water systems. Withdrawals from ground water are limited to those rates which will not cause: long term lowering of water levels; interference with the withdrawals of other permit holders; violation of water quality criteria; damage to the aquifer; or impact on the flow of perennial streams. Permits also require water conservation programs which are to include a leak detection element, a public awareness element, and encourage the use of water conservation devices.

Delaware's first septic system regulations were issued in 1968. While they established minimum isolation distances between septic system disposal areas and wells and minimum construction standards, it was not until 1985 that Regulations Governing the Design, Installation and Operation of On-site Wastewater Treatment and Disposal Systems were issued that required minimum separation distances between septic system disposal areas and ground water. These regulations are believed to be some of the most effective in the nation. They are based on the concept that septic tank effluent can be sufficiently renovated in the disposal field and underlying soil absorption area so that drinking water standards will be achieved at the site's property line. While this is not always possible for small lots that predate these regulations, new developments must meet the minimum siting density of not more than one single family residence per half acre or not more than 3,785 liters (1000 gallons) of septic tank effluent per acre per day. Additionally, septic systems must be sited such that a minimum of 91.44 centimeters (36 inches) of unsaturated soil is maintained between the bottom of the septic system disposal area and the seasonal high water table. Exceptions to that rule are for elevated sand mounds which require 111.75 centimeters (44 inches) and low pressure pipe systems (with small diameter pipe and shallow trenches) which require only 45.72 centimeters (18 inches). Approval, by the DNREC, of a site specific soils investigation is required for each site before an application for a septic system permit will be accepted by the DNREC. The investigation is conducted by a licensed soil scientist who determines the depth to the seasonal high water table based on two

chroma mottling. After the results of the investigation are approved by the

DNREC, the applicant has a designer prepare a permit application for the

appropriate septic system. Of the 3896 septic system permits issued by the

DNREC in 1990, 2973 were for gravity systems, 280 were for pressure dosed

systems, 291 were for low pressure pipe systems, and 293 were for elevated sand

mounds. The construction of 53% of the septic systems was inspected by the

DNREC staff to insure compliance with septic system construction standards and

permit conditions. Of the soils investigation reports submitted to the DNREC

since July 1985, 5.4% of the sites have been found to be unsuitable for any type

of septic system due to the seasonal high water table being less than 50.80

centimeters (20 inches) below the ground surface. Additionally, all soil

scientists, percolation testers, designers, and installers are licensed by the

DNREC.

Similar regulations have been issued to protect ground water from

contamination by the land application of sludge, the storage and land application

of manure, the spray irrigation of wastewater, the disposal of solid and hazardous

wastes, and underground storage tanks.

Wellhead and Aquifer Protection Program

The 1986 amendments to the Safe Drinking Water Act established a

requirement for the creation of State Wellhead Protection Programs (WHPP).

The purpose of a WHPP is to protect ground waters that supply public water

supply systems from potential contaminants. Delaware, with the DNREC as the lead agency, developed and is implementing a WHPP despite the lack of adequate federal funding.

Delaware's WHPP is composed of seven elements as follows:

o Specification of the Duties of State and Local Agencies - Development of a successful WHPP will require the coordination and assistance of many different agencies, organizations, and individuals. Therefore an advisory group representing state agencies, county and local governments, the University of Delaware, special interest groups, and environmental groups will be formed.

o Delineation of Wellhead Protection Areas (WHPAs) - This will include the development of criteria for delineation and the actual delineation of protection areas for over 400 public water supply wells.

o Contaminant Source Identification - Identification of all potential pollution sources within the defined wellhead areas (to be done in cooperation with well owners).

o Management Approaches - Development of approaches for use by local authorities to deal with pollution sources within WHPAs.

o Protection of New Wells/Wellfields - Development of methods which can be used by local authorities to protect new wells/wellfields from contamination.

o Contingency Plans - Development of general contingency plans for use in the event a water supply becomes contaminated. Plans would address temporary or alternate water supplies, long term water supplies, emergency contact, and financial funding.

o   Public Participation - Development of public informational programs to be presented to various civic groups and local public forums.

The DNREC submitted its Wellhead Protection Program to the United States Environmental Protection Agency (USEPA) and received approval on July 31, 1991. Legislation to amend the DEPA to enable the DNREC with specific authority to adopt and implement a WHPP, was introduced in the Delaware General Assembly in 1992, 1993, and again in 1994. While WHPP legislation has been supported by a small group of legislatures from the more populated areas of the State, the agricultural community has, so far, been successful at preventing its passage. The agricultural community has taken the position of opposing WHPP until provisions are made for them to be compensated for the taking of their lands that they perceive will occur when the delineation of large wellheads, especially around wells in the unconfined aquifer, restricts certain farming operations such as the storage of manure or pesticides and the application of certain pesticides.

Without specific enabling legislation, DNREC is working to implement its WHPP utilizing existing statutes and regulations and through coordination and cooperation with the Division of Public Health and local governments. Management of contamination sources in WHPAs will occur through zoning ordinances, site plan reviews, operating standards, source prohibitions, public education, and ground water quality monitoring.

Comprehensive State Ground Water Protection Program Profile

The DNREC in conjunction with the USEPA Region 3, completed the Delaware Comprehensive State Ground Water Protection Program Profile (CSGWPP) in September 1993. The CSGWPP provides an overview of the ground water related activities conducted in the State and describes the relationship of these programs to one another. The CSGWPP provides the basis for an assessment of Delaware's current ground water protection activities. The results of the assessment are being used to identify gaps in Delaware's ground water protection program where additional resources need to be targeted to complete the State's ground water protection efforts.

Public Water System Supervision

The Division of Public Health (DPH), Department of Health and Social Services, implements, under delegation from USEPA, the Public Water System Supervision (PWSS) Program in Delaware. While DNREC and DPH have worked closely on coordinating activities, certain common activities such as inspections of wells, creation of a common ground water data base, and issuance of certain permits have caused duplication of effort or gaps in monitoring and enforcement. In 1993 the Governor and General Assembly convened a Commission on Government Reorganization and Effectiveness. That Commission has recommended that the PWSS program be transferred from DPH to DNREC where it will become an integral part of DNREC's ground water protection program.   Implementation of the recommendation is planned for July 1994.

Conclusion

Ground water protection requires the consideration of economic, social, political, and environmental concerns and must balance those concerns to achieve the goal of protection of human health and the environment. Clean and plentiful ground water is essential to the continuance of quality of life in Delaware. Much has been done over the past decade to insure continued prosperity and environmental protection but there is a need to do more if future generations are to enjoy the same quality of life that we enjoy today.

Future ground water protection initiatives must include: pursuing the protection of existing or future sources of ground water, the active enforcement of rules and regulations for the protection of ground water quality and quantity, development of a geographic information system that is responsive to the interrelationship of ground water quality and quantity issues, adoption of a waste attenuation-well restriction policy, requiring in all environmental permitting programs the attainment of Delaware Drinking Water Standards at the point of contact with the water table, requiring agricultural activities to employ all feasible best management practices to protect ground water quality and quantity, and implementation of an aggressive program to educate the public and answer questions regarding ground water issues.

References

○ Delaware Comprehensive State Ground Water Protection Program Profile, DNREC and USEPA Region 3, September 30, 1993.

○ Delaware Environmental Protection Act, Delaware General Assembly, as amended through April 1993.

○ Delaware Regulations Governing the Construction of Water Wells, DNREC, January 20, 1987.

○ Regulations Governing the Design, Installation and Operation of On-site Wastewater Treatment and Disposal Systems, DNREC, as amended through August 10, 1987.

○ Regulations Governing the Allocation of Water, DNREC, October 2, 1986

○ State of Delaware, Groundwater Management Plan, staff of the DNREC, November 1, 1987.

○ The Management of Water Resources in Delaware, Summary Report, Comprehensive Water Resources Management Committee, July 1983.

WELLFIELD PROTECTION PROGRAM IN

BROWARD COUNTY, FLORIDA

Robert C. Shair[1] P.E.

Nazeer Ahmed[2] M. ASCE

INTRODUCTION

The Wellfield Protection Program in Broward County was
established because of the urgent need to protect existing and
future public utility potable water supply wells from the
irreversible and adverse effects of bacterial and chemical
contamination. The source of Broward County's drinking water is
the Biscayne Aquifer, a hydraulic unit composed of porous and
permeable limestones, sandstones and sand. In October, 1979, the
United States Environmental Protection Agency (EPA) designated
the Biscayne Aquifer a sole source aquifer. This action

---

[1] Retired. Formerly, Resources Development Manager, Water
     Resources Management Division, Office of Environmental
     Services, Broward County, Pompano Beach, Florida 33069

[2] Dean of Academic Affairs, Lenox Institute of Water Technology,
     P.O. Box 1639, Lenox, Massachusetts 01240

recognized both the importance of the Biscayne Aquifer as a water
supply and its vulnerability to contamination.

The Potable Water Wellfield Protection Program was initiated
early in 1981. It examined land uses in the County, made an
inventory of potential ground water pollution sources, compiled a
list of potential pollutants, delineated protection zones around
wells based on computer modeling of travel times, and developed a
protection strategy based on enforcement power legislation for
existing wellfields and land use controls to supplement the
legislation for future sites.

In December, 1983, approval was given by the Board of County
Commissioners to draft the legislation and promulgate the program.
After considerable public participation, an Ordinance and
Resolution were passed and became effective August 30, 1984.
Since that date, all activities in close proximity to potable
water supply wells have been permitted with controls, or have
been required to cease operation. Activities located further
away from wells, but still within Zones of Influence, are
regulated under permit.

AREA PROFILE AND TYPICAL WELLS

Broward County is situated close to the southern end of the
southeast coastline of Florida. The population today is
1 350 000 people. Underlying Broward County for developable area
of 1 062 km^2 is the Biscayne Aquifer, which is the County's sole
source of drinking water. The water is extracted from 318 wells
in 35 wellfields. The depth of the wells is approximately 30.5 m;
ground water level of the unconfined aquifer is 0.6 - 3.7 m below

the ground surface. The wells are currently permitted to pump 9.2 m³/s. The estimated population for the year 2020 is 1 900 000 persons having a projected water demand of 15.3 m³/s.

Potable water supply wells in Broward County are typically located in non-sensitive residential areas, but many existing wells have been located in areas where a threat of toxic material pollution exists. Examples of the location of potentially threatened wells in industrial areas are along railroad and highway right of ways, golf courses, at airports, on public work sites where machinery and vehicular maintenance is done, and near gas stations. Figures 1, 2, and 3 show typical wells.

Figure 1. Well on a Golf Course

Figure 2. Well on an Airport (also serving as an

end-of-runway marker)

Figure 3. Well near a Public Facility

ELEMENTS OF THE WELLFIELD PROTECTION PROGRAM

The elements of the program are listed in Table 1. For the
first element, the characteristics of the potential pollutants
were defined as listed in Table 2. They include toxic substances
or substances which have degradation products which are toxic,
substances which are prone to be persistent and substances which
are water soluble. All of these substances are listed by EPA,

TABLE 1 ELEMENTS OF WELLFIELD PROTECTION PROGRAM

- Identify wellfield pollutants and their sources
- Map zones of influence around wellfield
- Develop and implement strategies to minimize
  interaction between land uses and potable water
  wellfields

TABLE 2 POTENTIAL POLLUTANTS

- Toxic substances
- Toxic degradation products
- Prone to be persistent
- Water soluble

and the list includes the priority toxic pollutants. The list
also includes regulated substances that are covered by Florida
Statutes and the State Programs for the control of toxic and
hazardous materials. In the Wellfield Protection Program, there
are 192 listed regulated substances, which are confirmed by
consultants as being necessary to be controlled within Broward

County. Also, as part of this element there was compiled an
inventory of land uses and activities in the County which use
regulated substances. From it was developed a generic list of
sources of potential contamination as shown in Table 3. Some of

TABLE 3 SOURCES OF TOXIC AND HAZARDOUS WATER SUPPLY CONTAMINATION

. Fertilizers, pesticides and herbicides

. Urban runoff

. LUST

. Septic tanks

. Industrial activities

. Municipal services - wastewater treatment facilities -
  public works - landfills

. Land spreading and spray irrigation

the activities around the County that embody these potential
sources of contamination were close to potable water wells and
were subjected to action under the Wellfield Protection
Legislation.

For the second element, protective zones of influence around
wells and wellfields were defined, based on computer modeling of
travel times induced by the cone of depression around a well. All
of the modeling was done on wells that were pumped 0.044 m^3/s or
more. The average depths of water supply wells in Broward County
are about 30.5 m, and the drawdown can extend out for several
kilometers for larger wellfields. The drawdown creates a
hydraulic gradient toward the well mouth, and transport of

substances within the wellfield cone of depression will be
accelerated toward the well. Figure 4 shows a typical cone of
depression around a pumped well.

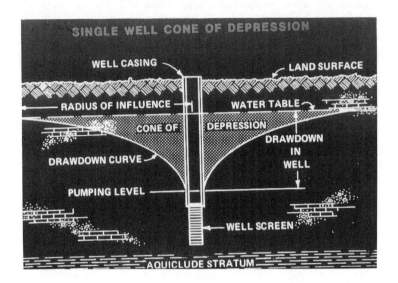

Figure 4   Single Well Cone of Depression

The Zones of Influence maps are based on travel time and
0.3 m drawdown contours. They were generated using a version of
the random-walk mass transport computer model of Prickett, Naymich
and Lonnquist of 1981. The method simulates pollutant movement
using particles released around wells in an inverted head/velocity
field. The head/velocity is calculated by using finite difference
computer modeling techniques that incorporate the effects of an
extensive canal system and year 2020 built-out pumping rates. The
pumping rates were determined by first projecting populating

figures for the year 2020 for each utility service area and multiplying this by a per capita consumption rate determined by the South Florida Water Management District (SFWMD). The Zones of Influence maps were finalized in January, 1984, and individual wellfield maps have been updated and adopted since that time. Revisions are due to changes in location of wells, location of new wells, changes in projected year 2020 pumpage rates and refinements of the modeling procedures. The Zones of Influence indicated on the maps are as follows:

(a) Zone 1: The land area situated between the well(s) and the ten-day travel time contour.

(b) Zone 2: The land area situated between the ten-day and thirty-day travel time contours.

(c) Zone 3: The land area situated between the thirty-day and the two hundred ten-day travel time contours, or the thirty-day and the 0.3 m drawdown contours, whichever is greater.

The travel time contour is defined as the locus of points from which water takes an equal amount of time to reach a given destination such as a well. The 0.3 m (1 foot) drawdown contour is defined by the locus of points around a well or wellfield where the free water elevation is lowered by 0.3 m (1 foot) due to the pumping of the well or wellfield. Typical Zones of Influence are depicted in Figure 5. Atypical Zone 1 is about 100 m in diameter. The 0.3 m drawdown for a large wellfield may be 1.6 km from the wells.

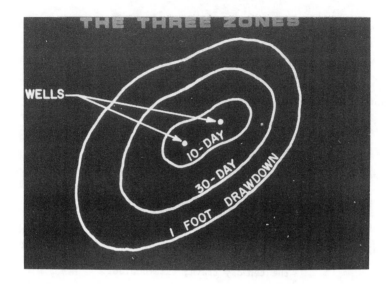

Figure 5.   Zones of Influence Around Two Wells

For the third element of the program, the Wellfield
Protection Legislation discussed below was developed and
implemented.

WELLFIELD PROTECTION LEGISLATION

The intent of the wellfield protection legislation
promulgated has been to safeguard the public health, safety, and
welfare by providing scientifically established criteria for the
prohibition and regulation of the storage, handling, use of
production of hazardous or toxic materials in close proximity of
wellfields. It was decided that the institution of enforcement
powers rather than land use restrictions was a better approach,

since use of specific substances could be prohibited or
controlled, even when they showed up in land uses which would be
impractical to restrict.

Wellfield Protection Ordinance 84-60 and Resolution 84-2025
were passed by the Broward County Commission on August 30, 1984.
They had first been made public in early 1984, and many public
meetings were held to debate the details. There were many
protests and expressed fears that the legislation would be
damaging to economic interests in the County, and great pains were
taken to listen to all groups and mold the legislation so as to
meet the intent of all concerned parties but minimize any specific
economic distress.

The legislation contains prohibitions for Zone 1 and
protection requirements for Zones 2 and 3. All of the provisions
apply to nonresidential activities, since most population risk is
associated with the quantities of materials used in nonresidential
activities. It was recognized that solvents, fertilizers, and
insecticides used in residential activities present some risk that
would have to be monitored and reviewed at a later date. The
cooperation of commercial lawn control services in residential
areas was pledged to apply chemicals in wellfield zones directly
in compliance with EPA container label restrictions and State of
Florida regulations. The bulk transport of chemicals into Zone 1
areas was restricted. Public education on wellfield protection
was a major element of the program. The intent of wellfield
protection has been to prevent contamination before it occurs,
and a monitoring program was put in place to detect trace

quantities of regulated substances in wells or in the water table before they reached the levels that would require shutting down the wells. The prohibitions and requirements for Zones 1, 2, and 3 are listed in Tables 4, 5, and 6.

TABLE 4 ZONE 1 PROHIBITIONS

- Nonresidential activities that store, handle, use or produce regulated substances shall cease in two years
- Ten specific exemptions and special exemption provision
- Administrative procedure provides for consideration for compensation

TABLE 5 ZONE 2 REQUIREMENTS

- Quarterly inventory of regulated substances
- Containment
- Daily surveillance
- Monitoring in raw drinking water; cessation mandated if accumulation occurs

TABLE 6 ZONE 3 REQUIREMENTS

- Quarterly inventory
- Daily monitoring for breakage or spillage
- In case of spillage, site reverts to Zone 2

ZONE 1 COMPENSATION

In Zone 1, there were fourteen activities that were either required to move or change their mode of operation. When the

Wellfield Protection Legislation was promulgated, it was
recognized that there would be economic consequences related to
these Zone 1 activities. The Broward County Commission decided
that protection of the wellfields was paramount, and that
provision should be made to consider the issues of compensating
these affected activities. It was initially estimated that
compensation costs could amount to 4-5 million dollars, but the
costs to close down contaminated wells and rebuild them elsewhere
was also estimated to be substantial. It was decided that
compensation for Zone 1 prohibitions was a county-wide water
supply issue, to be provided for from the general fund. Provision
was made for the staff to receive petitions for compensation, have
them reviewed by a Water Resources Advisory Committee, and then
submit them to the Board of County Commissioners for final
approval.

For four major cases, the prohibited activities were moved;
these included gas stations and municipal public works. There
were four other major cases involving municipally owned facilities
where Zone 1 interaction was eliminated by moving the wellfield.
Where municipalities were involved, there was an acknowledgement
on their part that a protected water supply was a Municipal as
well as a County responsibility and the total costs were fairly
shared. For the gas stations, settlements were negotiated with
the independent owners and petroleum companies to share the costs.
The final compensation cost amounted to $1.5 million, and all
parties agreed without any litigation or challenge to the
legislation.

CONCLUSION

Early in the 1980's, Broward County accepted the concept that
there was a need to protect its wellfields, and started a program
to define the issues and draft the necessary legislation.

The legislation first drafted early in 1984 was the subject
of many workshops and hearings, at which the apprehensions of all
affected parties were expressed. The apprehensions relating to
economic stress that might be caused were strongly voiced, but the
need to protect the sole source aquifer water supply were not
waived. Some compromises were made in the final legislation
adopted in August, 1984, but the original intent was preserved.
To protect the economic interests, a landmark provision was
included; to provide compensation where Zone 1 prohibitions
required an activity to relocate. The compensation provision
coupled with the willingness of municipalities to cost share, and
the reasonableness of affected industry to agree on a settlement,
all contributed to successful program. There were no challenges
or legal actions.

After ten years of operation, threatening activities have
been removed from very close proximity of wellfields, remaining
activities somewhat more distant have instituted rigorous
secondary containment and ground water monitoring, and a
considerable public awareness has been established.

A WATER SUPPLY PLAN FOR GROUND WATER MANAGEMENT IN HENDRY COUNTY, FLORIDA

Mark M. Wilsnack[1], Emily. E. Hopkins[2], and Amanda. J. Krupa[3]

[1]Senior Civil Engineer, [2]Senior Hydrogeologist and [3]Staff Hydrogeologist, Planning Department, South Florida Water Management District, 3301 Gun Club Road, West Palm Beach, Florida 33416-4680

INTRODUCTION

The South Florida Water Management District (SFWMD) undertook the preparation of a water supply plan for its Lower West Coast Planning Area in order to project future water demands, identify water sources and methods to meet this demand on a regional scale, and identify impacts to water resources that are associated with meeting these demands (SFWMD, 1994). The plan also includes recommended specific actions to be taken by the SFWMD in alleviating or minimizing undesirable impacts to resources.

The Lower West Coast Planning Area consists of Lee County, most of Collier County, and parts of Charlotte, Glades, Hendry, and Monroe Counties (Figure 1). The portion of this water supply plan dealing with Hendry County is presented in this paper.

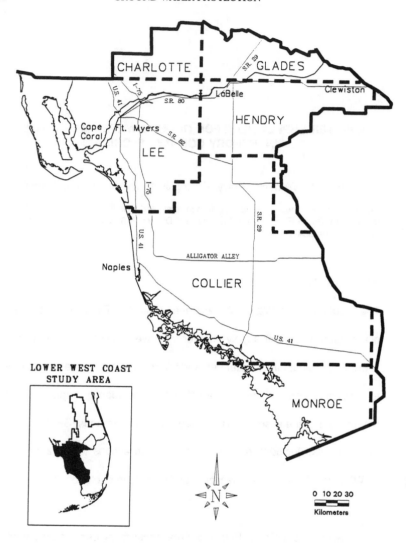

Figure 1. Lower West Coast Planning Area

The planning process followed in the development of the water supply plan for Hendry County can essentially be divided into three phases: (1) water resources assessment, (2) problem area identification through numerical model simulations, and (3) the investigation of potential solutions to the problem areas identified. The results of each of these planning phases are discussed in the following sections.

WATER RESOURCES ASSESSMENT OF HENDRY COUNTY

A detailed assessment of the hydrogeology of Hendry County and surrounding areas was performed by Smith and Adams, 1988. The study revealed three predominant freshwater aquifers. In descending order, these aquifers are the water table aquifer, the lower Tamiami aquifer, and the sandstone aquifer. The Tamiami confining zone separates the water table aquifer from the lower Tamiami aquifer while the upper Hawthorn confining zone separates the lower Tamiami aquifer from the sandstone aquifer. The geohydrologic properties of this aquifer system needed to construct the numerical ground water model were also determined in this phase of the study.

PROBLEM AREA IDENTIFICATION THROUGH NUMERICAL MODEL SIMULATIONS

Ground Water Use in Hendry County

Agricultural water use accounts for over 99% of the permitted water use in the county (Smith and Adams, 1988). Since pumpage records generally do not exist for agricultural uses, water use had to be estimated. Irrigation water use was estimated using the modified Blaney-Criddle method, as outlined by SFWMD 1993. Other types of agricultural water use were estimated using data from individual water use permits issued by SFWMD.

In order to determine the effects of ground water pumpage on the county's water resources using the resource protection criteria discussed below, two demand scenarios were simulated in the base numerical model runs described below: (1) a "presently permitted condition" scenario consisting of all 1990 permitted water withdrawals and (2) a future scenario consisting of projected 2010 ground water withdrawals. Since the primary cause for future increases in water demand in Hendry County is the conversion of pasture land to citrus and vegetables, 2010 ground water use projections were based on projected areas for these two crops.

Resource Constraints and Protection Criteria

Resource protection criteria define the extent to which hydrologic conditions can be altered without creating an unacceptable situation (e.g. aquifer dewatering, the alteration of a wetland hydroperiod, etc.). Commonly, such constraints are stated either as a maximum allowable drawdown or as a minimum allowable potentiometric surface elevation. In these cases, model simulated heads or drawdowns are compared to specified limits to determine whether or not the computed heads or drawdowns violate those limits. Problem areas are then identified by locating areas where the resource constraints are or will be violated. The two primary resource protection criteria imposed on the geohydrologic system of Hendry County pertained to the protection of wetlands and semi-confined aquifers.

According to the National Wetlands Inventory, there are about 19,000 hectares of wetlands in the portion of Hendry County located within the Lower West Coast Planning Area. These wetland communities are primarily forested and herbaceous systems. Drawdowns of the water table within these areas influence the vegetative species that comprise the area, which in turn influence the organisms that depend on the vegetation. In order to protect these environmentally sensitive areas, it is necessary to provide for their water needs through managing the water table aquifer. For planning purposes, the criterion developed by the Regulation Department of SFWMD for protecting wetlands from impacts due to pumping was imposed. This criterion is as follows: **water table drawdowns induced by pumping withdrawals should not exceed 0.3 meters for more than one month**

during any drought event that occurs more frequently than once every ten years in areas that are classified as wetlands.

The protection of semi-confined aquifers, on the other hand, is based on the principle that drawdown of the aquifer's potentiometric surface below a certain level will result in undesirable impacts to the aquifer. These impacts may include compaction, dewatering, reduced well yields, and upconing of saline water from deeper aquifers. Of these impacts, only aquifer dewatering was explicitly evaluated in this water supply plan.

For each of the three counties in the Lower East Coast Planning Area, the approach used to establish the minimum allowable potentiometric surface elevation for a semi-confined aquifer involved determining, through a kriging analysis of available data points, the variance in each model cell for the interpolated surface representing the top of the aquifer. All of this information was then used to determine an upper 90% confidence interval limit for the aquifer top elevation array based on the maximum variance for the areal extent of the aquifer that was pumped for water supply purposes. Using the surface representing the upper 90% confidence interval limit as the maximum allowable drawdown limit, as opposed to the surface representing the actual top elevation of the aquifer, provides a reasonable and defendable safety factor that is related to the uncertainty involved in establishing the top elevation of the aquifer. Unfortunately, a lack of data for the top elevation of the sandstone aquifer within Hendry County rendered this procedure unusable within the county. In order to circumvent this problem and address the uncertainty associated with the lack of data, the maximum variance computed for the most heavily pumped area of the sandstone aquifer within Lee County was used to establish the upper 90% confidence interval limit for the top of the sandstone aquifer within Hendry County.

A similar procedure was used to determine this confidence interval limit for the lower Tamiami aquifer within Hendry County; however, this procedure resulted in an unrealistically high potentiometric surface drawdown limit that would have required unusually high artesian pressures to be maintained in the aquifer. Consequently, an alternative approach had to be used to establish a buffer zone above the top of the aquifer. This alternate approach involved the comparison of aquifer top elevations with land surface elevations throughout areas of the aquifer where significant water use occurs. Based on these comparisons, a uniform value of 3 meters above the top of the aquifer was chosen as a reasonable buffer zone. Finally, in order to incorporate these buffer zones in the numerical model simulation runs described below, SFWMD imposed, for planning purposes, the following criterion: **the potentiometric surface should not decline below the selected, site specific level for any period of time during any drought event that occurs as frequently as once every ten years.**

Model Description

Background

The ground water system in Hendry County was modelled using the USGS three-dimensional ground water flow model, commonly known as MODFLOW (McDonald and Harbaugh, 1988). A brief description of this model is given below. For further details, see Smith, 1990.

Discretization

The ground water system in Hendry County was simulated in a quasi three-dimensional fashion using three layers, where layer 1 represents the water table aquifer, layer 2 represents the lower Tamiami aquifer, and layer 3 represents the sandstone aquifer. The model consists of 48 rows and 54 columns (figures 2 and 3).   Each of the cells is a square with dimensions of 1.61 km.

Boundary Conditions

In layers 1 and 2, the boundaries consist of constant head cells set 9.65 km outside of the county border except along Lake Okeechobee, where constant head cells were set along a canal which parallels the edge of the lake (figure 2). A buffer zone of 9.65 km was placed around the county border in order to minimize boundary effects (Smith, 1990). In layer 3, the boundary conditions are similar to those in layers 1 and 2 except for a no-flow boundary located at the easternmost extent of the sandstone aquifer (fig. 3).

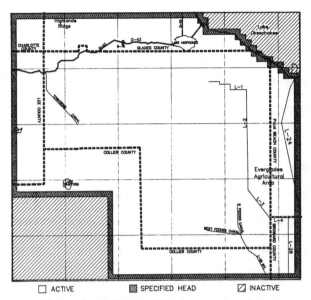

Figure 2. Cell Types for Layers 1 and 2

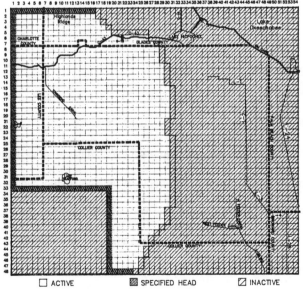

Figure 3. Cell Types for Layer 3

Stress Periods

The length of each stress period is one month. A stress period is, by definition, a period of time where the stresses on the geohydrologic system remain constant.

Recharge

Recharge to the model is derived from rainfall. For Hendry County, it was estimated that approximately 79% of monthly rainfall infiltrates into the soil profile (Smith, 1990). This portion of monthly rainfall was used as an estimate of monthly recharge to the water table aquifer. The monthly rainfall derived for each model cell and stress period is described in more detail below.

Evapotranspiration

In MODFLOW, the evapotranspiration (ET) rate is assumed to be linearly dependant on the potential ET (PET) rate, the ET surface, and the extinction depth. The ET surface is defined as the maximum water table depth where PET limits ET while the extinction depth is defined as the minimum water table depth where the maximum possible upward flux from the water table is equal to zero. Here, monthly PET values were estimated from pan evaporation data while the ET surface was taken to be the same as land surface. Extinction depths for each model cell were determined by Smith, 1990.

Surface Water Bodies

The river and drain modules of MODFLOW were used to simulate the interaction of the water table aquifer with surface water bodies. Only those surface water bodies with adequate, reliable data on cross sections, depths, and stages were accounted for in the model. These include two lakes and 15 canals (figure 2). For a more detailed description see Smith 1990.

Model Simulations

The numerical model simulation runs performed in conjunction with the Lower West Coast Water Supply Plan can be classified into two categories: "base case" runs and "alternative implementation" runs. The base case runs are essentially "do nothing different" runs, where both 1990-permitted and 2010 water users withdraw water as specified in their permits without being required to take any efforts to alleviate or avoid the types of undesirable impacts discussed previously. These base case runs incorporated the two water demand scenarios discussed previously with two different sets of rainfall conditions. The first set of rainfall conditions consisted of average monthly rainfall for Hendry County while the second set of rainfall conditions represented a historic 12-month drought event. This drought event was established by first ranking 50 historical, consecutive, 12-month rainfall events with respect to their monthly rainfall deficits below monthly rainfall conditions.

Next, a 12-month transient model simulation was performed using each of the top fifteen 12-month drought events. Simulated water table elevations were compared between model runs for key locations representing various land uses within Hendry County. The drought events were then subjectively re-ranked according to the magnitude and duration of the minimum water table elevations that were simulated at these locations. Finally, the fifth worst drought event (with respect to simulated water table elevations) was selected as an estimate of the 10-year, 12-month drought event. In the numerical model simulations, each set of rainfall conditions was incorporated with both demand scenarios to determine the chronic problems that will occur under average hydrologic conditions as well as the more acute impacts that will occur during drought conditions. As explained above, the latter type of impacts must be examined in light of the resource protection criterion.

In the alternative implementation simulations, the same water use demand and rainfall scenarios were simulated, but the model runs also incorporated practical measures for reducing the undesirable impacts in the base case simulations. By comparing the results of the two types of model simulations, it was possible to evaluate the effectiveness of proposed impact reduction measures under both the 1990 and projected 2010 water demands as well as both average and deficit rainfall conditions.

Model Simulation Results : Base Case

Figure 4 shows the results of the model simulations involving 1990 permitted demands under drought conditions. It was determined that the resource protection criterion was not met for 2784 hectares of wetlands, which is 14.3% of the total wetland area. For the projected 2010 water demands (figure 5), this number increased only slightly, to 3221 hectares, which is 16.4% of the total wetland area (it was determined during the course of this study that these estimates may be conservative and that they could be refined by making a number of enhancements to the model). Nonetheless, despite the small increase in wetland protection criterion violations between 1990-permitted and 2010 conditions, figure 4 reveals that there still may be an unacceptable, significant number of violations once (and if) all of the ground water appropriated in 1990 becomes utilized.

The aquifer protection criteria were only violated for the sandstone aquifer under 2010 water demands. These violations occurred over two moderately small regions located in the western portion of the county (figure 6). While these impacted regions of the aquifer are not aerially extensive, it is felt that they are significant enough to warrant concern over future aquifer dewatering. These results could also be refined in the future through model enhancements.

Frequency: 1 in 10 rain, 1990 base
Severity: >= 0.3 meters drawdown
Duration: > 1 month
Hectares: 2784 (14.3 percent of total hectares)

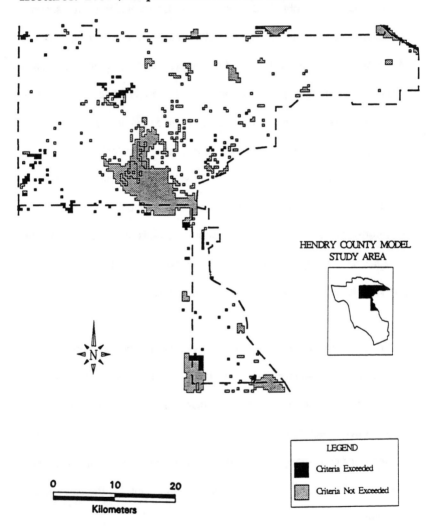

Figure 4. Wetland Areas With Criteria Exceeded
1990 Demands, Drought Conditions

Frequency: 1 in 10 rain, 1990 base
Severity: >= 0.3 meters drawdown
Duration: > 1 month
Hectares: 3221 (16.5 percent of total hectares)

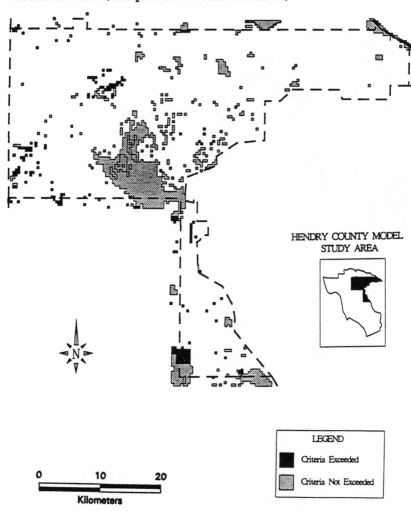

Figure 5. Wetland Areas With Criteria Exceeded
2010 Demands, Drought Conditions

Frequency: 1 in 10 rain, 1990 base
Severity: >= 0.3 meters drawdown
Duration: > 1 month
Hectares: 3108 (16.4 percent of total hectares)

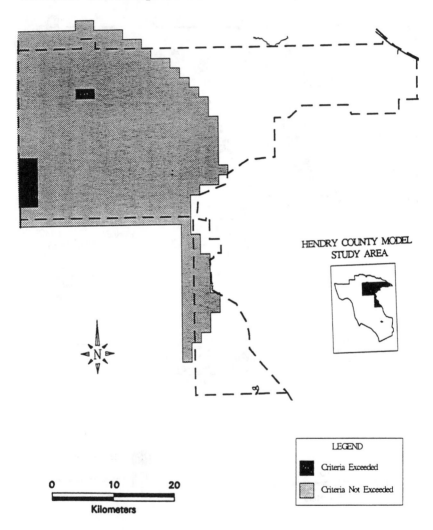

Figure 6. Aquifer Protection Areas With Criteria Exceeded
2010 Demands, Drought Conditions

PROPOSED SOLUTIONS AND IMPLEMENTATION ACTIONS

Model Simulation Results : Implementation of Alternatives

After considering a number of impact reduction measures, it was determined that only three water management alternatives could be readily applied. These were: (1) increase the irrigation efficiency for irrigation systems serving small vegetables to 75% for all users operating systems with efficiencies below this amount, (2) increase the irrigation efficiency for systems serving citrus to 85% for all users operating systems with efficiencies below this amount, and (3) implement both efficiency improvements.

Unfortunately, the improvements resulting from the application of the proposed alternatives in Hendry County turned out to be negligible. The total wetland area over which the resource protection criterion was not met remained nearly the same for both 1990-permitted and future water withdrawals. The same can be said for the total area over which the resource protection criterion was not met for the sandstone aquifer under 2010 water demands.

All of this implies that SFWMD needs to investigate the feasibility of implementing other solutions to the resource degradation problems identified

in this study. The remaining alternative solutions that presently appear viable
are discussed briefly below. For more detailed discussions, the reader is
referred to (SFWMD 1994).

Other Viable Solutions and Implementation Actions

Develop New Surface Water Sources

There are numerous surface water courses in Hendry County, ranging from
field scale drainage systems to major canals, including the Caloosahatchee
River (figure 2). During the wet season, these water courses act primarily as
drainage systems, removing large amounts of fresh water which are
eventually discharged to tide. The feasibility of capturing a portion of this
fresh water at both the field and regional scales and storing it in either
agricultural reservoirs or in deeper aquifers via ASR (Aquifer Storage and
Recovery) should be investigated. The availability of this extra water during
the dry season would help to reduce the stresses imposed on the three
aquifers studied in this water supply plan.

Develop New Ground Water Sources

The use of deeper aquifers may also be a viable alternative for reducing the
demands placed on the shallower aquifers within Hendry County.   In
particular, the Floridan Aquifer System, a deep, semi-confined aquifer

separated from the sandstone and lower Tamiami aquifers by a thick semi-confining unit, is known to exist regionally throughout the Lower West Coast Planning Area. In order to assess the feasibility of using this aquifer as a source of irrigation water in Hendry County, additional information on the geohydrologic and geochemical properties of the aquifer must be obtained. The SFWMD is currently in the process of obtaining this information.

Implementing Water Shortage Triggers

More efficient management of the ground water resources of Hendry County could be attained if management strategies for drought periods were directly incorporated into the permitting and allocation process. Such a strategy could include the use of "drought triggers" or "water shortage triggers". In this case, a water shortage trigger would be a key water level that would initiate or trigger management actions (such as water use cutbacks) by either SFWMD or the permittee. These triggers could initially be based on the resource protection criteria discussed earlier and could be revised as the resource protection criteria are refined to become more site and resource specific.

APPENDIX 1. REFERENCES

McDonald, M. G. and Harbaugh, A. W. (1988). "A Modular Three-Dimensional Finite Difference Ground-Water Flow Model." <u>Techniques of Water Resources Investigations of the United States Geological Survey,   Book 6, Chapter A1.</u> U.S. Government Printing Office, Washington, D.C.

Smith, K. R. (1990). "A Three-Dimensional Finite Difference Ground Water Flow Model of Hendry County." <u>Technical Publication 90-04</u>, South Florida Water Management District, West Palm Beach, Florida.

Smith, K. R. and Adams, K. M. (1988). "Ground Water Resource Assessment of Hendry County, Florida: Part 1 - Text." <u>Technical Publication 88-12</u>, South Florida Water Management District, West Palm Beach, Florida.

South Florida Water Management District, (1994). <u>Lower West Coast Water Supply Plan</u>, South Florida Water Management District, West Palm Beach, Florida.

South Florida Water Management District, (1993). <u>Basis of Review</u> Permit <u>Information Manual, Volume III</u>, SFWMD, West Palm Beach, Florida.

GROUND WATER PROTECTION IN ARIZONA

T.W. ANDERSON[1], M. ASCE

Introduction

Archeological evidence indicates that development of water resources in Arizona began as early as 300 B.C. when the Hohokam Indian culture developed an agricultural-based society and diverted waters of the Salt and Gila Rivers through a network of canals to irrigate land adjacent to the rivers (Haury, 1976). In the middle to late 1880's, early non-Indian settlers used available surface waters to support agricultural development. Because of the erratic nature of streamflow in parts of Arizona, ground-water resources were soon developed to supplement the highly variable surface-water supply. Increased agricultural and industrial development was accompanied by increased demand on ground-water systems. This development, in turn, resulted in the need to protect ground-water supplies from long-term degradation of both quantity and quality. As a result of recent State legislation, Arizona has developed one of the most comprehensive ground-water protection programs in the United States.

Background

Protection and management of water resources by Federal, State, and local agencies was not a concern during the early stages of development in Arizona. Early ground-water laws in Arizona established the right of the land owner to reasonable and beneficial

---

[1] Senior Hydrologist, Errol L. Montgomery & Associates, Inc., 1075 E. Ft. Lowell Rd., Suite B, Tucson, Arizona, 85719.

use of percolating ground water that occurred under the land; State courts decided that such ground water was not subject to appropriation (Leshy and Berlanger, 1988). In the early to middle 1900's, development and use of water resources was encouraged through the enactment of Federal and State programs and subsidies and, as a result, overdevelopment of water resources in some parts of the State occurred.

Non-Indian development of ground-water supplies, chiefly for irrigation use, began in Arizona about the turn of the century. In 1915, about 152 million cubic meters (123,000 acre-feet) of ground water was withdrawn from wells in Arizona (U.S. Geological Survey, 1982). Ground water withdrawals in the State totaled 1.23 billion cubic meters (1 million acre-feet) in 1940, and averaged about 6.17 billion cubic meters (5 million acre-feet) per year during the 1970's and 1980's. During and after the 1930's, withdrawals greatly exceeded the rate of natural recharge to ground-water systems; this imbalance in the hydrologic system is termed "overdraft". The results of overdrafting the ground-water systems were decline in ground-water levels and partial depletion of aquifer storage. Total ground-water withdrawals from the turn of the century through 1980 were about 228 billion cubic meters (185 million acre-feet); Anderson and others (1990) estimated that the net overdraft that resulted from these withdrawals was nearly 123 billion cubic meters (100 million acre-feet).

Arizona's earliest attempts to resolve the overdraft problem was through passage of the Groundwater Code of 1945. The code was a well-registration bill and did little to lessen the rapidity of ground-water development. The 1945 code was followed by the Groundwater Code of 1948, which provided for designation of critical ground-water areas within which there could be no further expansion of agricultural acreage. The 1948 code prohibited drilling of new wells to irrigate land not previously irrigated. However, the code did not limit the amount of pumping from existing or replacement wells nor did it restrict non-irrigation uses. In 1977, the State legislature established the Groundwater Management Study Commission. The Commission was charged with the responsibility to develop a comprehensive ground-water management code for Arizona.

Arizona Groundwater Management Act of 1980

The Arizona Groundwater Management Act of 1980 was the result of nearly 3 years of education, discussion, and decision making by members of the Arizona Groundwater Management Study Commission (Arizona Groundwater Management Study Commission, 1980). The Act was designed to reduce long-term overdevelopment and to bring the overdrafted ground-water basins back into hydrologic balance by the year 2025. Historically, agriculture was the driving force in the rapid development of use of ground water; in 1980 irrigated agriculture used 82 percent of the ground-water withdrawal in the State (U.S. Geological Survey, 1982). As a result of enactment of the Arizona Groundwater Management Act of 1980, use of water for agriculture was to be systematically reduced and financial constraints were to be used to encourage efficiency in water use.

The Groundwater Management Act of 1980 has been called "a landmark in the history of United States water management" (Morrison, 1983). The act established four Active Management Areas (AMA) that encompass areas having the greatest depletion in aquifer storage, and established two Irrigation Nonexpansion Areas (INA) that encompass areas having lesser development problems; subsequently, another INA has been established in the Harquahala Plain area. The location and extent of the AMAs and INAs are shown on **Figure 1**. Approximately 80 percent of Arizona's population resides within the AMAs, and about 70 percent of the State's annual ground-water consumption occurs within the AMAs.

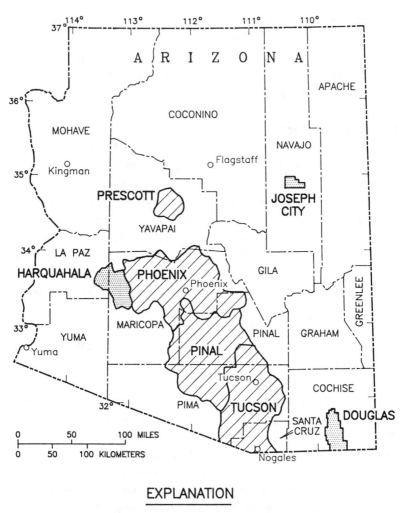

EXPLANATION

▨ ACTIVE MANAGEMENT AREA

▨ IRRIGATION NONEXPANSION AREA

FIGURE 1. ACTIVE MANAGEMENT AREAS AND IRRIGATION
NONEXPANSION AREAS IN ARIZONA

The Groundwater Management Act of 1980 provides four methods for the Arizona Department of Water Resources (ADWR) to manage ground-water resources within the AMAs. The methods include: 1) establishment of a system of grandfathered ground-water rights, which restricted withdrawals to the amount of the grandfathered right and prohibited ground-water withdrawal by those not holding a right; 2) requirements for conservation by all ground-water users; 3) allowance for purchase and retirement of irrigated agricultural lands; and 4) provision for augmentation of water supplies.

In the Phoenix, Prescott, and Tucson AMAs **(Figure 1)** the management goal was to reach a condition of 'safe yield' by the year 2025. Safe yield was defined as a long-term balance of withdrawal and recharge. In the Pinal AMA, the management goal was to preserve the existing agricultural economy as long as possible while being consistent with long-term needs to preserve future water supplies for other uses. Plans for ground-water management were to be implemented over five management periods beginning in 1980 and ending in 2025; the plans would include four 10-year plans and one plan for the final 5-year period. Water conservation goals become increasingly stringent in each subsequent period.

The first management plans, starting in 1980, required moderate mandatory conservation to achieve reductions in ground-water uses by 1987. Different conservation requirements were mandated for each of the designated water-use sectors of the four AMAs. Agricultural users were required to implement moderate conservation measures in the first management plan; moderate conservation was to be achieved by assignment of an annual allocation or agricultural water duty, and a goal of 55- to 60-percent farm efficiency. Municipal water providers were required to meet a percent reduction of their 1980 water-use rates. The second management plans started in 1990 and assigned, in most cases, a reduced water duty for agricultural users and a goal of 85-percent farm efficiency (Barrios, 1989). Municipal water providers were assigned an individual conservation requirement based on each municipal providers potential for conservation. The second management plans also contained management strategy for augmenting the AMA water supplies. Augmentation strategies include optimizing use of Central Arizona Project water, watershed management, weather modification, reuse of effluent, use of urban runoff, and maximization of recharge potential.

The Groundwater Management Act of 1980 also provides for ADWR to levy a withdrawal fee on all groundwater withdrawn from an AMA. The fee may not exceed $5.00 per 1,234 cubic meters (1 acre-foot) of withdrawal; revenues from the fee will be used to finance administrative and enforcement expenses, augmentation of water supplies, and purchase and retirement of grandfathered rights beginning January 1, 2006.

The Groundwater Management Act of 1980 created a comprehensive system of well regulations, including registration requirements, permit requirements, approval requirements for drilling in certain areas, and licensing of well drillers. The statutes also include regulations for closure of wells; the regulations are designed to protect against contamination of ground water. Examples are: wells penetrating single or multiple aquifer systems with vertical flow components must be plugged in a manner to prevent fluid communication between aquifers, and no materials containing organic or toxic substances may be used in abandonment of a well.

Arizona Environmental Quality Act of 1986

In Arizona, existing and potential ground-water quality problems are a major concern because ground water is the principal source for public supply. Efforts to develop a ground-water quality protection program began in the late 1970's and continued into the 1980's. The Arizona Department of Environmental Quality (ADEQ) was established by the Arizona Environmental Quality Act of 1986 to administer State water-quality programs for both surface water and ground water, as well as other environmental programs. ADEQ is responsible for recommendation and adoption of water-quality standards, definition of the boundaries of all aquifers in the State, development of permit programs for aquifer protection and underground injection control, administration of the Water Quality Assurance Revolving Fund (the State superfund program), and water monitoring.

In setting aquifer water-quality standards, ADEQ must consider: 1) protection of public health and the environment; 2) uses that have been and may be made of the aquifers; 3) requirements of the Clean Water Act and Safe Drinking Water Act; and 4) any guidelines, action levels, or numerical criteria adopted or recommended by the U.S. Environmental Protection Agency (EPA). Primary drinking water standards established by EPA are required to be adopted by ADEQ; in addition, ADEQ has authority to establish numerical standards, if warranted to protect human health, for constituents for which EPA does not have a standard.

An Aquifer Protection Permit (APP) must be obtained by any person who discharges or who owns or operates a facility that discharges pollutants. Discharge is broadly defined to include the direct or indirect addition of a pollutant to the waters of the State; discharge from a facility could occur either directly to an aquifer, or to the land surface or the vadose zone in such a manner that there is a reasonable probability that a pollutant will reach an aquifer (Arizona Revised Statutes § 49-201[10], 1986). To obtain an APP, an applicant must demonstrate that the facility will be designed for maximum discharge reduction, and that any discharge would not cause a violation of an aquifer water-quality standard at an applicable point of compliance. The permit process requires detailed hydrogeologic studies, technical review by ADEQ, and a public meeting and hearing before the decision to grant or deny the permit is made.

Summary

Protection of ground water is being accomplished chiefly through a management approach in Arizona. Water supply quantity and water-quality management strategies rely on different laws and are implemented by separate State agencies. The available volume of ground water in storage in the future will be nearly the same as at present if the long-term management goal of safe yield is met in the AMAs by the year 2025. Ground-water quality is being protected chiefly through the permitting process, which is applicable to any facility that might discharge pollutants to the land surface, the vadose zone, or the aquifer. Water-quality standards applicable to ground water have been established by EPA and ADEQ for most constituents in drinking water.

References

Anderson, T.W., Freethey, G.W., and Tucci P., 1990, **Geohydrology and water resources of alluvial basins in south-central Arizona and parts of adjacent states:** U.S. Geological Survey Open-File Report 89-378, 99 p.

Arizona Groundwater Management Study Commission, 1980, **Final report:** Phoenix, Arizona Groundwater Management Study Commission report, 49 p.

Arizona Revised Statutes, 1986, **Title 49, The Environment:** Chapter 2, Article 1, section 201, paragraph 10.

Barrios, F.M., 1989, **Arizona's second management plan:** in Harris, S.C., ed., Proceedings of the 16th Annual Conference, Water Resources Planning and Management Division, American Society of Civil Engineers, Sacramento, California, pgs. 614-617.

Haury, E.W., 1976, **The Hohokam--Desert farmers and craftsmen:** Tucson, University of Arizona Press, 412 p.

Leshy, J.D., and Belanger, J., 1988, **Arizona law where ground and surface water meet:** Arizona State Law Journal, Vol. 20, pgs. 657-748.

Morrison, Allen, 1983, **Arizona's water strategy--Bring more in and restrict its use:** American Society of Civil Engineers, Civil Engineering, V. 53, No. 4, pgs. 46-49.

U.S. Geological Survey, 1982, **Annual summary of ground-water conditions in Arizona, spring 1980 to spring 1981:** U.S. Geological Survey Open-File Report 82-368, 2 sheets.

UTAH GROUND WATER QUALITY PROTECTION STRATEGY

by Carl H. Carpenter[1], P.E., F.ASCE
and Larry Mize[2], P.E.

INTRODUCTION

Utah's water pollution control program is segregated in to three areas: surface water, ground water, and underground injection. The surface and ground water programs and part of the underground injection program are administered by the Water Quality Board through the Division of Water Quality in the Department of Environmental Quality. A portion of the underground injection control program is administered by the Division of Oil, Gas and Mining, a subdivision of the Utah Department of Natural Resources. A separate program for the protection of drinking water supplies is administered by the Drinking Water Board through the Division of Drinking Water in the Department of Environmental Quality. Water well construction, monitor well construction, and well abandonment are administered through the Utah Division of Water Rights of the Department of Natural Resources. (Tudermann, 1993)

[1]Principal Engineer, Provo City Department of Water Resources, Provo, Utah

[2]Manager, Ground Water Protection Section, Utah Division of Water Quality, Department of Environmental Quality, Salt Lake City, Utah

of Environmental Quality. Water well construction, monitor well construction, and well abandonment are administered through the Utah Division of Water Rights of the Department of Natural Resources. (Tudermann, 1993)

REGULATION OF DISCHARGES TO GROUND WATER

Ground Water Protection Regulations, designated as Utah Administrative Code R317-6, were adopted in August 1989, and are intended to protect the existing and potential future beneficial uses of ground water by protecting the quality of ground waters in the state. (DEQ, April 1993) Through the use of stringent "protection levels", the regulations focus on prevention, rather than cleanup, of ground water contamination. The regulations reflect an antidegredation policy that recognizes that there will be some effect on ground water by anthropogenic activity, but seeks to minimize these effects to acceptable levels. In keeping with the goal of protecting beneficial uses, the level of protection afforded ground water varies with the quality and expected uses of the groundwater such that higher quality ground water, e.g. drinking water, is accorded greater protection than say a source for irrigation. Protection levels for specific aquifers are established through a classification process that may be initiated either by the State or by local jurisdictions. Compliance is administered through a permit system applicable first to new facilities and, as deemed necessary by the Executive Secretary to the Board of Water Quality, to existing facilities or those that have been permitted by rule.

Any facility or activity which causes or may cause a discharge of pollutants to ground water is required to obtain a ground water discharge permit. These include, but are not limited to, land application of wastes, storage of wastes, large feedlots, mining, milling, and metalurgical extraction operations, including heap leach facilities, and wastewater and process water impoundments. Some facilities and activities may qualify for "permit by rule" and do not require going though the formal permitting process. Examples of "permit by rule" include facilities or activities that are regulated by other agencies (such as coal mines regulated by the Division of Oil, Gas and Mining) or where the activity has a negligible impact on ground water. In these cases, however, the facility may still be responsible for any ground water contamination it may cause. (Barnes & Croft, 1986)

GROUND WATER QUALITY STANDARDS

The ground water quality regulations establish seven classes of ground water:

Class IA - Pristine Ground Water

Class IB - Irreplaceable Ground Water

Class IC - Ecologically Important Ground Water

Class II - Drinking Water Quality Ground Water

Class III - Limited Use Ground Water

Class IV - Saline Ground Water

Classes are based upon the ambient aquifer water quality which in turn determines the use that can be made of the water. Water quality is protected by two related measures-- ground water quality standards and ground water class protection levels. (DEQ, April 1993)

The ground water quality standards are numerical standards for potential ground water contaminants. These standards are based on Maximum Contaminant Levels (MCLs) established by the State and EPA for drinking water. The standards are not applied directly but are used as reference levels, and protection levels are set as percentages of the ground water quality standards. For total dissolved solids (TDS), the protection level is based on the background value for each class, while the standards for physical characteristics and radionuclides equal the protection level.

Classes IA, IB, II, as their names suggest, include the highest quality ground water, including ground water that can be used for drinking with conventional treatment. Class IC ground water is protected as a source of water for potentially affected wildlife habitat. Class III ground water is protected for agricultural and industrial uses and as a potential source of drinking water with "substantial treatment."

In no case is a concentration of a contaminant in Classes IA, IB, II and III allowed to exceed the ground water quality standards adopted by the regulations. Class IC limits are determined on a case-by-case basis to meet appropriate surface water standards. Protection Levels for Class IV ground water are also established on a case-by-case basis to protect human health and the environment. Unclassified ground water is protected to a level consistent with existing ground water quality.

GROUND WATER DISCHARGE PERMIT PROGRAM, (DEQ, April 1993)

General Provisions

The Ground Water Quality Protection Regulations are intended to protect existing and probable future beneficial uses of Utah's ground water resources. The level of protection afforded ground water varies with the quality of the water such that higher quality ground water is accorded a greater degree of protection. Compliance is administered through a permit system applicable first to "new facilities" and, as deemed necessary by the Executive Secretary, to "existing" and "permit by rule" facilities.

Application Requirements

The owner or operator of a "new facility" that "discharges or would probably result in a discharge of pollutants that may move directly or indirectly into ground water", must apply to the Executive Secretary for a ground water permit before constructing, modifying, installing or operating the facility.

The regulations identify twenty-one types of sources that are granted a permit by rule. A facility permitted by rule is not required to obtain a discharge permit or to comply with any other provision of the regulations, except that such a facility may not cause any ground water to exceed the ground water quality standards or the applicable class TDS limits. However, the Executive Secretary may require a facility permitted by rule to submit a ground water permit application if the Executive Secretary determines that discharge from the facility may be causing or is likely to cause exceedances of the ground water quality standards or applicable class TDS limits, or otherwise is interfering or may interfere with the probable future beneficial use of the ground water.

The regulations establish seventeen types of information that generally must be supplied with a permit application. The required information includes technical information about the facility and the area where it is located; technical information about the type and nature of the discharge; a proposed monitoring plan, including a sampling plan; for facilities located in areas of unclassified ground water, information concerning the quality or receiving ground water; and for existing sources, a corrective action plan to correct any violations of ground water quality standards that existed prior to issuance of the permit.

A discharge permit for a new facility requires the facility to use "best available treatment and methods" to minimize discharges. For Class III ground waters, the Board may approve an alternate concentration limit for a new facility under limited conditions.

A discharge permit for an existing facility shall require that the facility use treatment and discharge minimization technology commensurate with plant process design capability and similar or equivalent to that utilized by facilities that produce similar products or services with similar production process technology. An alternate concentration limit may be approved by the Board for an existing facility under limited conditions.

Before issuing a permit, the Board must publish a Notice of Intent to approve the Permit. Interested persons are allowed thirty days in which to comment to the Board on the proposed permit.

<u>Term, Termination and Transfer</u>

Discharge permits are issued for five year periods and renewed or extended for additional periods not to exceed five years. Permits may be terminated by the Executive Secretary for a variety of reasons, including a determination that the permitted facility endangers human health or the environment and can only be regulated to acceptable levels by permit modification or termination. Discharge permits may be transferred with a permitted facility by giving timely notice to the Executive Secretary.

If monitoring discloses that a facility is not complying with discharge permit requirements the Executive Secretary may require accelerated monitoring, a plan to assess the source, extent and potential dispersion of contamination and potential remedial action, and immediate corrective action. The Board intends to develop cleanup standards for ground water in the near future. In the interim, cleanup standards will be set on a case-by-case basis.

Discharge permits require compliance monitoring and reporting. Compliance monitoring points must be as close as possible to the discharge point and, absent written agreement of the affected property owners, the compliance monitoring points must not be beyond the boundaries of the facility. In no event shall a compliance monitoring point be located within the radius of influence of any beneficial use water supply.

Permit Appeals

Any person who is denied a permit by rule or who objects to a discharge limit or to other ground water permit conditions or requirements made by the Executive Secretary may request a hearing before the Board. Any person whose ground water permit is denied or proposed to be terminated or revoked may appeal that decision to the Executive Director of the Department of Environmental Quality.

UNDERGROUND INJECTION CONTROL PROGRAM, (DEQ, 1991)

Under Utah law, underground injection means "the subsurface emplacement of fluids by well injection," and underground injection is prohibited unless authorized by permit or regulations. An underground injection permit does not authorize any activity that endangers a public drinking water source or may adversely affect human health.

Under Utah Administrative Code R317-7 injection wells are classified into five categories. Class I includes the well injection of hazardous, industrial, and municipal wastes beneath the lowermost formation within two miles of the well bore containing an underground source of drinking water ("USDW"). Class II wells inject fluids brought to the surface in connection with certain oil, gas, and other hydrocarbon production and storage operations. Class III wells inject materials for the extraction of minerals such as sulfur and in situ uranium production. Class IV wells include injection of hazardous or radioactive wastes into or above a formation which, within two miles of the well, contains a USDW. Class V wells consist of a variety of other injection wells including drainage and recharge wells as well as multiple dwelling cesspools.

Class IV wells are prohibited unless allowed under EPA regulations. Class I, III, and V wells are regulated by the Board's Executive Secretary. Class II wells are regulated by the Division of Oil, Gas and Mining. Unless specifically exempted, the owner or operator of a new or existing Class I, III, or V underground injection well must obtain a permit from the Executive Secretary.

The Executive Secretary may issue permits on an area, rather than an individual, basis and has broad discretion in issuing such general permits. Precise permit procedures and conditions are further detailed in the regulations. When the operator and owner are different individuals it is the operator's duty to obtain a permit. Information required to complete the application includes mapping, proposed operating data, and other specifics of the particular well system. Permits for Class I and Class V wells are effective for no longer than ten years. Permits for Class III wells are issued for a period up to the operating life of the facility, but are reviewed at least every five years. Transfer, modification, and termination of permits, as well as identification of conditions waiving requirements, are all controlled by the Executive Secretary as outlined in the regulations. The regulations also contain detailed technical requirements covering matters such as well construction, operation, plugging and abandonment.

Class II injection wells involve conventional oil and gas production techniques, such as recovery of oil and natural gas and storage of liquid hydrocarbons. The requirements for Class II injection wells parallel those of other well categories. Such wells must be maintained in a manner to avoid pollution to any USDW and must confine injected fluids to the area approved. Applications for approval of Class II wells are submitted to the Board of Oil, Gas and Mining and must be accompanied by a variety of specific geological and mechanical data. Notice and hearing requirements are also specified in the Class II injection well regulations.

An order approving a Class II injection well is valid for the life of the well unless revoked by the Board of Oil, Gas and Mining. That Board may revoke or administratively amend a Class II injection well permit for just cause and after notice and hearing. Under certain conditions it may exempt an aquifer from USDW classification, but any such decision is subject to EPA approval.

DRINKING WATER SOURCE PROTECTION PROGRAM (DEQ, July 1993)

In July 1993, the Drinking Water Source Protection Rule was implemented by the State Drinking Water Board designated as Utah Administrative Code R309-113. This authority governs the protection of ground water sources of drinking water, and it applies to all public water supplies (PWS) with ground water sources of drinking water. Each public water supply is required to prepare and submit a source protection plan to the Division of Drinking Water (DDW) which includes a delineation report; an inventory of potential sources of ground water contamination; a management program to control each contamination source; a plan for controlling or prohibiting new potential contamination sources; an implementation schedule, and a contingency plan.

Delineation-Three management zones are delineated so that more intensive management can be focused in the most vulnerable areas. Drinking Water Source Protection (DWSP) zones are delineated around each ground-water source according to the following criteria and thresholds:

- ▸ Zone One is a 100-foot fixed radius around the well or margin of the spring called the "accident prevention zone." Its purpose is to prevent accidents and to protect the annulus of the well or the spring collection area.

- ▸ Zone Two is a 250-day time of travel to the well or spring called "attenuation zone." Its purpose is to reduce concentrations of pathogenic microorganisms and some chemicals to levels below maximum contaminant levels (MCLs) before ground water reaches the well or spring.

▸ Zone Three is a 15-year time of travel to the well or spring called the "remedial action zone." Its purpose to provide protection to the drinking water source and to afford sufficient time for remediation or developing a new source in case of a contamination incident.

An optional, two-mile radius delineation procedure may be used to establish protection areas. This procedure is most appropriate for small, rural systems whose ground-water sources are located in areas where there are few if any potential contamination sources. Source Protection Plans include:

▸ Inventory-An inventory of potential contamination sources within each of the DWSP zones is required. Each potential contamination source identified on the inventory is then ranked according to the risk it poses to a particular well or spring.

▸ Management program for each potential contamination source- A management program for each potential contamination source is developed to prevent contamination of the public drinking water source.

▸ Plan for controlling and prohibiting new potential contamination sources- A plan is required to prohibit or control any new contamination sources that request to locate within established DWSP zones.

▸ New wells and springs- Prior to constructing any new ground-water sources of drinking water each PWS is required to develop a Preliminary Evaluation Report which demonstrates that a source meets certain requirements. PWSs are required to submit Preliminary Evaluation Reports and Engineering Plans and Specifications to DDW concurrently; review by DDW is also conducted concurrently. DDW will not grant plan approval to a PWS until requirements are met. Construction standards relating to protection zones and management areas (fencing, diversion channels, sewer lines, etc.) are found in R309-106. (DEQ, 1979) After the source is constructed a DWSP Plan shall be developed, submitted, and implemented accordingly.

▸ Minimum management requirements for new wells and springs- PWSs are required to exclude pollution sources from Zone One in all aquifer settings and Zone Two in unconfined aquifer settings. Examples of pollution sources include, but are not limited to, the following: storage facilities that store the liquid forms of extremely hazardous substances, septic tanks, drain fields, Class V underground injection wells, landfills, open dumps, landfilling of sludge and septage, manure piles, salt piles, pit privies, drain lines, sewer lines, and animal feeding operations with more than ten animal units.

- Contingency plans- A contingency plan for the entire water system is also required from each PWS.

OTHER GROUND WATER PROTECTION PROGRAMS

UNDERGROUND STORAGE TANK PROGRAM, (DEQ, 1989)

Utah enacted the Underground Storage Tank Act in 1989, establishing the underground storage tank ("UST") program and authorizing the Solid and Hazardous Waste Control Board to adopt implementing regulations. The UST program applies to the same tanks regulated under Subtitle I of RCRA and adopts by reference EPA UST technical standards and financial responsibility requirements. The state program also establishes requirements and procedures not found in the federal program, and is governed by Administrative Code R311. The UST program includes the testing, monitoring, and remediation procedures for underground storage tanks.

WATER WELL CONSTRUCTION AND ABANDONMENT, (DNR, 1987)

The construction and abandonment of water wells is administered by the Division of Water Rights thru Administrative Rules for Water Well Drillers, Utah Code 73-3-25. The rules govern the licensing of well drillers, proper construction standards for water wells (except geothermal wells) monitor wells, and proper abandonment of wells. Geothermal well construction is administered by the Division of Oil, Gas and Mining.

REFERENCES

1. Tundermann, David W., Utah Environmental Law Handbook, Government Institutes, Inc. 1993.

2. Barnes, Robert P., and Croft, Mack G., Ground Water Quality Protection Strategy for the State of Utah, Utah Department of Health, 1986.

3. Utah Department of Environmental Quality, Division of Drinking Water, Drinking Water Source Protection Rule, R309-113, July 1993.

4. Utah Department of Environmental Quality, (DEQ), Division of Water Quality, Administrative Rules for Ground Water Quality Protection, R317-6, April 1993.

5. Utah Department of Environmental Quality, (DEQ), Division of Water Quality, Administrative Rules for the Underground Injection Control Program, R317-7,   May 1991.

6. Utah Department of Natural Resources, (DNR), Division of Water Rights, Administrative Rules for Water Well Drillers, Utah Code 73-3-25, July 1987.

7. Utah Department of Environmental Quality, (DEQ), Division of Environmental Response and Recommendation, Administrative Rules for Underground Storage Tanks, R311-200, 1989.

8. Utah Department of Enivornmental Quality, (DEQ), Division of Drinking Water, Rules for Public Drinking Water Systems, Part II, Design and Construction Standards for Systems, R309-105 through 112, October 1979.

GROUND WATER MANAGEMENT IN THE SALT LAKE VALLEY
SALT LAKE COUNTY, UTAH

Michele M. Lemieux and Jerry D. Olds[1]

INTRODUCTION

In managing ground water basins, the Utah Division of Water Rights attempts to limit ground-water withdrawals to the annual recharge to prevent aquifer over-drafting. Typically, the major consequence of over-drafting an aquifer is the lowering of the potentiometric surface, making it more difficult and costly to use the water. However, in the Salt Lake Valley, excessive withdrawals from the primary aquifer could threaten the quality and usability of the entire aquifer.

Now, the Salt Lake Valley is facing a major problem: the ground water rights are over-appropriated. Recent studies indicate that current ground water withdrawals are approaching the safe annual yield of the principal aquifer.

---

[1]Associate Engineer, Assistant State Engineer, Utah Division of Water Rights, 1636 West North Temple, Salt Lake City, Utah 84116-3156.

222

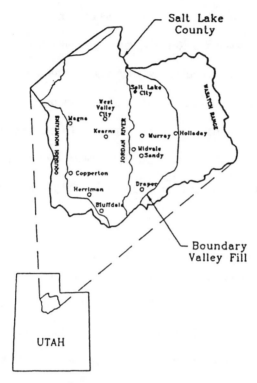

Figure 1. Location Map of Salt Lake Valley
Ground Water Basin

However, there are many approved water rights in the valley that have not yet been developed. If these water rights were to be developed, the additional withdrawals could allow poor quality water to move into the high quality regions of the aquifer, reducing and perhaps even destroying the usability of many portions of the aquifer. To address the over-appropriation of the valley, the Utah Division of Water Rights implemented a Ground-Water Management Plan. A location map of the Salt Lake Valley is shown in Figure 1.

BACKGROUND

There are numerous studies on the geology, hydrology, and contaminant susceptibility of the Salt Lake Valley (Baskin, 1990; Waddel and others, 1987a and 1987b; Hely et. al., 1971). Therefore, this discussion will be limited to the basics necessary to understand the management of water rights in the valley.

The Salt Lake ground water basin consists of a deep, unconsolidated aquifer, overlain by a relatively thin confining layer, which is in turn overlaid by a shallow aquifer. The confining layer is not continuous and does not extend to the edges of the valley fill, thus near the mountain fronts the principal aquifer is unconfined. The water quality of the deeper aquifer ranges from excellent on the eastern side of the valley to poor on the west.

The water quality of the shallow aquifer is generally poor. The deep or principal aquifer is the main source of ground water for the valley. The shallow aquifer is currently not used extensively.

Recharge to the principal aquifer primarily comes from subsurface inflow from fractured consolidated rock. The shallow aquifer is recharged by infiltration from streams, precipitation, irrigation and by upward leakage from the principal aquifer. Where the confining layer is absent, infiltration will also recharge the principal aquifer. The quality of recharge water varies singificantly throughout the valley, with the highest quality water from the Wasatch mountains to the east.

There is an upward gradient from the principal to the shallow aquifer over a large percentage of the valley, which helps maintain the high quality of the principal aquifer. Evidence indicates, however, that excessive pumping in the principal aquifer can reverse the upward gradient, allowing downward leakage of the poor quality water. This has indeed happened in the past; Waddell and others (1987b) go over a specific example in detail. There are several portions of the principal aquifer that are susceptible to contamination if the hydraulic gradient becomes reversed for a sufficient length of time (Baskin, 1990).

THE LIMITED RESOURCE PROBLEMS

It is necessary to estimate the safe annual yield of the Salt Lake Valley to determine if restriction on withdrawals are necessary. The safe yield of the principal aquifer is assumed to be equivalent to the amount of recharge the principal aquifer receives annually. However, it is difficult to utilize a valley wide recharge estimate, primarily due to the size, geography and complexity of the basin. For the most part, the benches on the western and eastern portions of the valley are not hydraulically connected. In addition, there are extensive variations in water quality. Therefore, the safe annual yield must be estimated for specific areas of the principal aquifer, rather than the entire valley. Recharge to areas of similar water quality were calculated using recharge components input into the Salt Lake Valley flow model (Waddell and others, 1987a).

*East-Wasatch Mountains (High Quality)	$1.51 \cdot 10^8 m^3/yr$
*Southeast (High Quality)	$2.96 \cdot 10^8 m^3/yr$
*West-Oquirrh Mountains (High Quality)	$2.10 \cdot 10^7 m^3/yr$
*Northwest (Low Quality)	$4.07 \cdot 10^7 m^3/yr$
*South (Low Quality)	$2.47 \cdot 10^7 m^3/yr$
*West-Oquirrh Mountains (Low Quality)	$3.95 \cdot 10^7 m^3/yr$

Based on these figures, the total recharge to the principal aquifer is about $2.79 \cdot 10^8 m^3/yr$

To determine the extent of over-appropriation, the ground water withdrawals must also be determined. However, only a portion of the approved and perfected water rights have been developed. Thus, not only do current ground water withdrawals need to be estimated, but POTENTIAL ground water withdrawals as well. In other words, the probable withdrawal by both undeveloped rights, and rights for which water use information is not available. Work is on-going to to estimate the potential withdrawals under existing water rights for the recharge areas defined above, in order to determine the extent of over-appropriation on a more local basis. Potential withdrawals for the entire Salt Lake Valley according to beneficial use is estimated to be $4.94 \cdot 10^8 m^3/yr$ (Utah Division of Water Rights, 1991). Typically, the amount of water approved under a water right is far greater than what is actually used, so the potential withdrawals could be much higher. Thus, if unrestricted development of water rights is allowed, ground water withdrawals could exceed recharge, with catastrophic results.

The concern here is not solely a water supply issue; how excess withdrawals will effect the water quality becomes equally, if not more, important. This and other considerations need to be evaluated to determine the safe annual yield from the aquifer. As discussed previously, there is more than enough evidence to indicate that a reversal of the vertical hydraulic gradient causing downward movement of poor quality water into the prinicpal aquifer can easily happen. Maintaining the upward gradient is not the only water quality issue; a substantial portion of the principal aquifer on the western side of the valley is contaminated from early mining practices (Waddell and others, 1987b). Excess withdrawals in the surrounding areas could potentially impact the ground water flow regime and cause these contaminants to spread to more pristine regions of the aquifer.

The serious nature of the potential to overdraft the aquifer and adversely affect the water quality required immediate action. Work was initiated to determine the current withdrawals from the principal aquifer. To further define how withdrawals can effect the ground-water flow regime, a study was started by the United States Geological Survey (USGS) in cooperation with the Utah Division of Water Rights, Utah Division of Water Quality, and the majority of public water suppliers in the valley.

Restrictions on water rights were also immediately necessary. Under Utah water law, the State Engineer must distribute and deliver water under the doctrine of prior appropriation. The appropriation doctrine requires that junior rights be shut off if necessary in order to satisfy senior rights. However, when applying this doctrine to a ground-water basin, there are some problems and unresolved issues. For instance, should a 1972 priority water right diverting poor quality water on the west side of the valley be shut down because of a 1942 water right diverting high quality water from the east? What about later priority rights that have already drilled and developed their wells; can they be shut down by early priority rights that are still in the process of developing their well and perfecting their right? What about water right transfers?

Obviously, additional quidelines are needed in order to administer the water rights in the Salt Lake Valley. In response, the Division of Water Rights implemented the "Interim Ground Water Management Plan" for the Salt Lake Valley in April of 1991. "Interim" refers to the current USGS study; when the study is complete, the results will be used to define a final version of the plan. The remainder of this paper will discuss the details of the management plan.

INTERIM GROUND WATER MANAGEMENT PLAN--SALT LAKE VALLEY

Withdrawals

Ground water use in the Salt Lake Valley will be administered according to the doctrine of prior appropriation. The quantity of water that can be withdrawn from the principal aquifer will be limited, and water rights cut according to priority date. Restrictions on withdrawals will be based upon the cumulative effects in different areas of the valley. Thus, what are referred to as "managment areas" are defined. The "management area" boundaries generally coincide with variations in the water quality of the prinicpal aquifer. by maintaining safe annual yield within each management area, it is hoped to prevent migration of poor quality water into high quality areas. To even further protect the high quality regions of the aquifer, "buffer zones" are defined. Figure 2 illustrates the management areas in the valley and the respective withdrawal limits of each area.

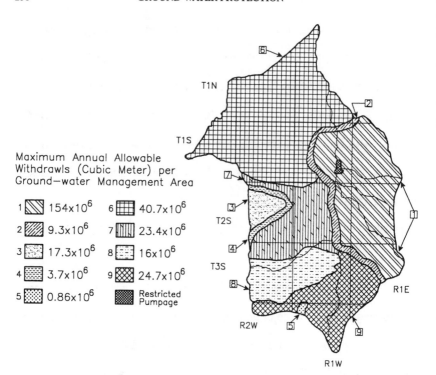

Figure 2. Proposed distribution of ground water withdrawls in Salt Lake Valley, principal aquifer.

For each area considered to be of generally poor quality, the maximum withdrawal is equal to the total recharge. The poorer quality water in the southwestern portion of the valley is further divided into two regions, in order to account for high sulfate ground water believed to be attributed to early mining activity. For each high quality area, slightly different criteria, based on the percent of high quality recharge each area receives, is used to distribute the withdrawals. The maximum withdrawal allowed in the buffer zones is reduced in half on a unit area basis with respect to the areas they are designed to protect. Where the shallow aquifer is suspected to be heavily contaminated, withdrawals in the deeper aquifer are completely restricted.

The management plan has one other limitation to withdrawals based entirely on water quality. If excessive withdrawals in a particular area are adversely effecting the overall water quality of that area, diversions will be required to cease. Under this scenario, water rights within the affected area would be administered on priority.

Metering and Sampling

To aid the Division of Water Rights in monitoring withdrawals in each management area, part of the management plan requires all water users that pump more than $6.20 \cdot {}^4 m^3$/yr to install a meter on their wells and submit monthly withdrawal data. If a water user pumps more than $3.1 \cdot 10^5 m^3$/yr, an annual water quality sample is also required. This will help the Division in determining if withdrawals in one area are beginning to adversely effect the overall water quality.

Applications to Appropriate

The Salt Lake Valley principal aquifer is closed to new applications except for single family use. Under the condition when a public water system is available, the water right will be abandoned.

Water Right Extensions, Transfers, and Submittal of Proof

Extensions of time will be critically reviewed. If the applicant is not diligently pursuing development, the extension will either be denied, the priority date reduced or quantity of water reduced. On the same note, when proof is submitted, a much stricter position will be taken by the Division. The water right will be limited to the amount of water that has been placed to beneficial use at the time of proof submittal. All water right transfers will be evaulated based on their potentials to adversely effect water quality and/or exceed maximum withdrawals. Transfers from the shallow aquifer to the deeper aquifer, or from a poor quality water area to a high quality area will not be approved.

Miscellaneous Issues

Wells will be regulated so that no more than 3.7m of drawdown may occur on another well with an earlier priority date. Water users in a particular area may enter into an agreement to provide a variance from this requirement.

During years of drought, there will be more demand on the ground water. Therefore, in the final version of the management plan, to be implemented after the completion of the USGS study, management area withdrawals will be evaluated over a five year moving average to make better conjuctive use of surface and ground water.

CONCLUSIONS

Recent studies indicate that the ground water in the Salt Lake Valley is greatly over-appropriated. Excessive withdrawals have the potential to not only mine the aquifer, but also to adversely effect the ground water quality. The interim ground water management plan is intended to guide future development of Salt Lake Valley's ground water resources in a fair and responsible manner, while still protecting senior rights in accordance to the doctrine of prior appropriation. Both water quantity and water quality issues are addressed in the management plan, primarily by restricting withdrawals.

REFERENCES

Baskin, R. L., 1990 Selected Factors Related to the Potential for Contamination of the Prinicpal Aquifer, Salt Lake Valley, Utah, USGS Water Resources Investigations Report 90-4110

Hely, A.G., Mower, R.W., and Harr., C.A., 1971, Water Resources of Salt Lake County, Utah, Utah Department of Natural Resources, Division of Water Rights, Technical Publication No. 31

Waddell, K. M., Seiler, R. L., Santini, M. and Solomon, D. K., 1987a, Ground-Water Conditions in Salt Lake Valley, Utah, 1969-83, and Predicted Effects of Increased Withdrawals form Wells: Utah Department of Natural Resources, Division of Water Rights Technical Publication No. 87

Waddell, K. M., Seiler, R. L., and Solomon, D. K., 1987b, Chemical Quality of Ground Water in Salt Lake Valley, Utah, 1969-85, and Predicted Effects of Increased Withdrawals from Wells: Utah Department of Natural Resources, Division of Water Rights Technical Publication No. 89

Utah Division of Water Rights, 1991, Salt Lake Valley Ground Water Priority Lists

California's Ground Water Quality Protection Strategy

Wendy L. Cohen, P.E., M.ASCE[1]

Introduction

California's State Water Resources Control Board (State Board) and the nine Regional Water Quality Control Boards (Regional Boards) are responsible for protecting the quality of the waters of the State for present and future beneficial uses. The Boards were created by the 1949 Dickey Act, and in 1969, the Porter-Cologne Water Quality Control Act greatly increased and broadened the Boards' authority to protect water quality (1).

Beneficial uses of water include municipal and domestic supply, industrial supply, agricultural supply, navigation, recreation, hydroelectric power generation, ground water recharge, freshwater replenishment, and preservation and enhancement of fish, wildlife and other aquatic resources. Drinking water supply generally is the most sensitive beneficial use of water. However, some other beneficial uses require more stringent limits for some chemicals, such as boron in agricultural supply water and copper in water used for fishery resources.

State Board Resolution No. 88-63, the *Sources of Drinking Water Policy,* designates all surface and ground water in the state as suitable or potentially suitable for municipal or domestic water supply with some exceptions. For example, a water source would not be considered suitable as drinking water if the total dissolved solids exceed 3,000 mg/l, the sustained yield of a well is less than 200 gallons per day, or the water has contamination which cannot reasonably be treated using best available technology.

---

[1] Senior Water Resources Control Engineer, California Regional Water Quality Control Board, Central Valley Region, 3443 Routier Road, Sacramento, CA 95827

Ground water is a vital resource in California. California has some 450 ground
water basins with a combined storage capacity of 1.3 billion acre-feet which underlie
about 40 percent of the land surface. Of the total water used in the state, 40 percent
comes from ground water including:

> 46 percent of public drinking water supplies
>
> 93 percent of rural drinking water supplies
>
> 54 percent of industrial supplies
>
> 39 percent of agricultural supplies

Many past waste disposal practices have contaminated shallow ground water and
the deeper aquifers from which water supplies are drawn. Examples of such
practices are pesticide and fertilizer applications to agricultural land, toxic waste
disposal in unlined ponds, spills and leaks at industrial facilities due to poor operating
practices, spills at bulk fuel terminals, unmanaged dairy waste disposal, leaking
underground storage tanks, and solid waste disposal in unlined landfills.

The goals of California's ground water quality protection strategy are to maintain
ground water quality to fulfill present and future beneficial uses and to restore the
quality of ground water where feasible and appropriate. Actions to achieve these
goals include establishing standards and programs to prevent pollutants from entering
the ground water, maintaining existing high quality ground water, and restoring
polluted aquifers where technically and economically feasible. A number of programs
are designed to meet these goals, such as regulation and permitting of waste
discharges to land, testing and design requirements for underground and
aboveground petroleum storage tanks, pesticide labelling requirements and use
restrictions, and several cleanup programs.

Ground Water Quality Protection Programs

Waste Discharge Requirements

Porter-Cologne requires waste dischargers to file a report of any proposed waste
discharge (RWD) which could affect water quality. Each RWD must be accompanied
by the appropriate annual fee which is based on the discharge's threat to water quality
(fees range from $200 to $10,000). Based on the RWD and other available
information, the appropriate Regional Board holds a public hearing and prescribes
waste discharge requirements (WDRs). Most WDRs also contain monitoring and

reporting programs which the dischargers must follow on a regular basis to demonstrate compliance with the WDRs.

If a discharger violates WDRs, the Board can issue an "order to cease and desist" instructing the discharger to stop violating or threatening to violate WDRs (1). The Board also can issue an order to "clean up ... or abate the effects" of waste discharged in violation of WDRs or in such a manner that a condition of pollution or nuisance is created or threatened. The Cleanup and Abatement Order (CAO) also can be issued by the Board's Executive Officer without formal Board action. If a discharger fails to comply with these orders, the Executive Officer or the Board can impose fines administratively, or refer the case to the State Attorney General's office for civil and/or criminal proceedings.

Land Discharge Regulations

Additional requirements to those above apply when waste is discharged to land. Section 13172 of Porter-Cologne gave the State Board authority to "classify wastes according to the risk of impairment to water quality" and "to classify the types of disposal sites according to the level of protection provided for water quality" (1). This statewide disposal site and waste classification system is contained in the California Code of Regulations (CCR), Title 23, Division 3, Chapter 15 entitled *Discharges of Waste to Land*. The purpose of the regulations is to prevent pollution and unacceptable water quality degradation resulting from discharges of liquid and solid waste to land. In 1984, the State Board overhauled the Chapter 15 regulations and amended them again in 1993 to be consistent with Subtitle D of the federal Resource Conservation and Recovery Act.

The Chapter 15 regulations establish minimum requirements for waste discharge to land to protect water quality, including waste and site classifications, facility siting criteria, detailed construction standards, mandatory water quality monitoring provisions, corrective action requirements, and strict closure and post-closure maintenance standards. Types of facilities covered include surface impoundments, landfills, land treatment facilities, and waste piles.

Underground Storage Tank Regulations

In 1982, the Legislature added Chapter 6.7 to Division 20 of the Health and Safety Code to regulate underground storage tanks. This law, along with the associated *Underground Storage Tank Regulations* (CCR Title 23, Division 3, Chapter 16), requires local agencies to permit, inspect, and oversee monitoring programs to detect leakage from underground storage tanks. The local agency or the Regional Board directs the cleanup of contaminated soil and ground water resulting from a leak or other unauthorized discharge from the tank or appurtenant piping. In any case, the various agencies will coordinate to ensure that requirements from each agency are consistent.

Aboveground Petroleum Storage Act

In 1989, the Legislature passed the Aboveground Petroleum Storage Act (APSA) in response to a large spill from a Martinez refinery into the Suisun Marsh, a sensitive wetland in the San Francisco Bay Delta (2). Subject to the law are owners or operators of a facility operating aboveground tanks greater than 660 gallons capacity (or cumulative capacity greater than 1,320 gallons), who must file a storage statement and filing fee with the State Board every two years. The storage statement lists the number of tanks, and their contents and capacities. The APSA also requires each aboveground tank facility to prepare and implement a Spill Prevention Control and Countermeasure (SPCC) Plan. The SPCC plan has been required by federal law since 1975 (40 CFR Part 112).

The Regional Boards use the fees generated by the APSA to inspect tank facilities and review their SPCC plans and secondary containment devices. If an inspector finds evidence of a spill or leak that could impair water quality, the Board can require installation of a ground water monitoring well network. Once contamination is discovered, cleanup is required as discussed below.

Pesticide Regulations

The State Department of Pesticide Regulation (DPR) has adopted rules for seven chemicals found in ground water due to agricultural use of currently registered pesticides (3). These chemicals are listed on DPR's Ground Water Protection List as "detected leachers" and include atrazine, simazine, bromacil, diuron, prometon, bentazon, and aldicarb (CCR Title 3, Chapter 4, Subchapter 1, Article 1). Their use is restricted to specific areas at specific times, and the user must obtain a permit. Eleven other pesticides are listed as "suspected leachers" on the Ground Water Protection List based on their physical and chemical characteristics and on their use patterns.

These pesticides are subject to dealer sales reporting and to monitoring by DPR. DPR has proposed an additional 38 pesticides to be listed as "suspected leachers."

Ground Water Cleanup Strategy

When ground water has become contaminated, it must be cleaned up to restore its beneficial uses. The party who caused the pollution is responsible for its remediation. However, if this person is not able to pay for cleanup, the Regional Board can require the property owner to undertake the necessary measures. If an operator or property owner does not comply with the actions required in letters from staff, staff can use one of Porter-Cologne's enforcement options, usually a CAO. The CAO contains a compliance time schedule for specific tasks, including investigation of the extent of contamination and a remediation plan to clean up the ground water to its previous quality.

For remediation of contaminated sites, the Regional Board develops soil and ground water cleanup levels to protect water quality. Soil cleanup levels are based on the leachable concentration of the constituent(s) of concern (COCs) in the soil for non-volatile compounds or on the concentration in soil vapor of volatile COCs; the depth to ground water; and the potential attenuation of the COCs in the soil column. For constituents which have already reached ground water, the soil must be cleaned up to a level such that no additional ground water degradation will occur.

The ground water cleanup goal is set at the level of each COC in nearby uncontaminated ground water, called the background level. Therefore, during the ground water investigation, the responsible party (RP) must define the extent of the contamination to these background levels. For most organic chemicals, the background level is zero or, in reality, the lowest level detectable by laboratory analytical methods. In the feasibility study of alternative cleanup methodologies, the RP must evaluate the alternative of cleaning up ground water to background levels, including its technical and economic feasibility. If the best available technology cannot technically or economically achieve background levels, the Regional Board will re-evaluate and possibly modify the cleanup goal. However, the final cleanup goal will not exceed that necessary to protect beneficial uses.

Progress toward cleaning up contaminated ground water is slow, but is proceeding more quickly at the military bases and other industrial facilities than in agricultural areas where ground water is polluted with nitrates and the pesticide,

DBCP (the use of DBCP was outlawed more than 20 years ago). Another area of little progress is water supplies contaminated with the solvent perchloroethylene due to discharges of dry cleaning wastewater to sewer lines in several Central Valley cities.

Conclusion

Ground water is used extensively in California as a major water supply for people, agriculture and industry. Ground water contamination has occurred over the years due to various practices. A number of water quality protection programs initiated in the 1970s and 1980s have imposed new requirements so that businesses operating now will not cause similar pollution. These laws also require cleanup of the past problems.

References

1. California Water Code, Division 7, Section 13000 et seq., *Porter-Cologne Water Quality Control Act,* 1969

2. Health and Safety Code, Division 20, Section 25270 et seq., *Aboveground Petroleum Storage Act,* 1989

3. Food and Agriculture Code, Section 13141 et seq., *Pesticide Contamination Prevention Act,* 1985

Ground Water Impacts from Drainage Disposal

Dale K. Hoffman-Floerke[1]
Charyce L. Taylor[2]

Introduction

When the California State Water Project was
authorized in 1960, the Department of Water Resources was
authorized to investigate and develop drainage options
for water users in the State service area. In response
to construction of evaporation ponds, which provide an
interim method of disposing of subsurface agricultural
drainage, the Department of Water Resources established
the Evaporation Pond Investigation in August 1986.

The investigation's primary objectives are:

1. Identify and document potential environmental
    effects resulting from evaporation ponds.

2. Identify, evaluate, and demonstrate alternative
    pond design and management methods to reduce the
    size and potential adverse impacts of the
    evaporation ponds.

3. Recommend and demonstrate improvements to
    existing guidelines for design, construction,
    and operation of the evaporation ponds in the
    most environmentally safe and cost-effective
    manner.

----

[1]Recreation and Wildlife Resources Advisor, California
Department of Water Resources, 3251 "S" Street,
Sacramento, CA 95816.
[2]Environmental Specialist II, California Department of
Water Resources, 3374 East Shields Avenue, Fresno, CA
93726.

4. Work with regulatory agencies and operators of active evaporation ponds that are in compliance with the waste discharge requirements to coordinate monitoring efforts.

5. Disseminate current information regarding evaporation ponds to pond developers, operators, and regulatory agencies.

Current Regulatory Status

The California Department of Toxic Substance Control and the California Regional Water Quality Control Boards regulate evaporation ponds under several California statutes. The Regional Boards have the primary responsibility to regulate the ponds under the Porter-Cologne Water Quality Control Act to protect the beneficial uses of the State's water (California Department of Water Resources, 1988). On August 6, 1993, the California Central Valley Regional Water Quality Control Board certified environmental impact reports prepared for the active evaporation ponds and adopted waste discharge requirements for all of the ponds. Five of these ponds are slated for closure within the next 12 months (California Central Valley Regional Water Quality Control Board, 1993a). The Board demands that the pond operators comply with the waste discharge requirements in order to protect the beneficial uses of State water.

In establishing waste discharge requirements, the Board considered the requirements of the California Environmental Quality Act and determined whether environmental impacts will occur as a result of the construction and operation of each pond. The waste discharge requirements will be periodically updated by the Board, and reports summarizing environmental impact monitoring will be submitted and reviewed every two years.

Title 22 of the California Code of Regulations classifies and regulates waste water according to the potential hazard it poses to public health, wildlife, and other components of the environment. While primary responsibility for regulation of hazardous waste rests with the California Department of Toxic Substance Control, the California Regional Water Quality Control Boards may enforce the Toxic Pits Cleanup Act of 1984 if a pond's chemical constituents exceed hazardous levels. Additional regulatory requirements apply if the pond water has been classified as hazardous to human health. A pond must be closed if such a hazard exists.

Evaporation Pond Locations

The majority of the evaporation ponds used for drainage water disposal are located in the southern part of the San Joaquin Valley; a natural outlet for surface drainage from the Valley does not exist. The agricultural lands in the northern part of the Valley are allowed to drain into the San Joaquin River as long as water quality objectives for the river are met. Because discharge into the river is an option in the northern part of the Valley, use of evaporation ponds is generally unnecessary there. However, in order to protect water quality in the San Joaquin River, measures to control timing of discharges are under consideration.

Evaporation Pond Characteristics

There are currently 22 evaporation ponds in the southern San Joaquin Valley ranging in size from 3 to 740 hectares. The total area of the active ponds is about 2,235 hectares. Approximately 24,000 hectares of farmland drain into the ponds.

The evaporation ponds range from small, single-celled units to larger, multi-celled systems. Some ponds have windbreaks or islands which were originally constructed for erosion control. Because evidence shows that these features encourage bird use of the ponds, removal of windbreaks and islands is stipulated in all of the waste discharge requirements for active ponds. Although most of the ponds are not lined, Westlake Farms, Inc., is lining the cell embankments of its evaporation ponds as part of its mitigation for impacts to wildlife (California Central Valley Regional Water Quality Control Board, 1993c).

The average life expectancy of an active evaporation pond is 40 years. This life expectancy depends on the management of the salts that accumulate during operation of the pond.

Environmental Concerns

Construction and operation of the evaporation ponds have raised numerous environmental concerns including concerns about surface and ground water contamination.

The potential for seepage of evaporation pond drainage to degrade ground and surface water has been recognized since ponds were first constructed in the Valley. Seepage from poorly constructed ponds could degrade the quality of shallow ground water near the ponds and the quality of surface water down-gradient

from the ponds. The California Central Valley Regional
Water Quality Control Board, concerned about the
potential for degradation of ground water, made the
determination that, for all ponds, any impacts on shallow
ground water would be localized and would not affect any
beneficial uses. Furthermore, although the deeper
confined ground water (beneath the "E" clay) is generally
of good quality, because of the low permeability of the
soils underlying the ponds, minimal migration of the
shallow ground water to the deeper confined aquifer will
occur. All of the waste discharge requirements are
consistent with the antidegradation provisions of the
State Water Resources Control Board's Resolution 68-16.
Operator compliance with the waste discharge requirements
will render any impacts to ground or surface water
quality insignificant (California Central Valley Regional
Water Quality Control Board, 1993a).

Regulations for management of evaporation ponds
have been specifically designed to limit the facilities'
potential impacts on ground and surface water to
acceptable levels. The Tulare Lake Basin Water Quality
Control Plan (approved by the State Water Resources
Control Board) indicates that the ponds do not have
a significant adverse impact on usable ground water
(defined as water having total dissolved solids less than
10,000 parts per million or milligrams per liter) if pond
water is contained by a clay layer at least 1 meter thick
with a hydraulic conductivity of less than 10^{-10} meters
per second. If the ground water is not usable, clay with
a hydraulic conductivity of 10^{-8} meters per second is
considered adequate. Interceptor drains are generally
considered adequate to protect ground and surface water
adjacent to ponds having lateral seepage.

Drainage evaporation ponds in the Valley usually
overlie areas having shallow layers of heavy clay soil
and shallow, relatively poor-quality ground water.
These characteristics contributed to the drainage
problems that necessitated construction of evaporation
ponds. Available monitoring data on drainage inflows
to the ponds and on shallow ground water near the ponds
indicate that all of the active ponds overlie unusable
shallow ground water (California Central Valley Regional
Water Quality Control Board, 1993a).

Available information on the near-surface geology in
the Valley suggests that existing ponds are situated in
areas where one or more thick layers of low-permeability
clay lie within 6 meters of land surface. However,
detailed studies of near-surface geology at specific
sites suggest that, although the ponds are situated in
areas of generally low permeability, many of them may be

underlain by pockets of some relatively permeable materials. The heterogeneity of bottom materials makes it difficult to predict average seepage rates from the ponds (McCullough-Sanden, 1986) and has raised concerns that some ponds may have more vertical seepage than regional near-surface geological studies originally predicted.

Lateral seepage has been a particular concern in many areas because the clay layers often have permeable sands sandwiched between them. Provisions of the waste discharge requirements stipulate that all active ponds must have interceptor drains to contain such seepage.

Determination of a pond's impacts on ground and surface water is sometimes difficult because pond seepage may gradually decline. While infiltration rates are governed by complex processes and can vary significantly from one area to another, numerous studies have demonstrated that soils can become less permeable after inundation. Several factors including the presence of microorganisms, secretions from microorganisms, deposition of clay particles, and deposition of precipitated salts probably contribute to reducing the hydraulic conductivity of inundated soils (California Central Valley Regional Water Quality Control Board, 1988).

To date, the California Central Valley Regional Water Quality Control Board has based its evaluation of potential pond seepage on the hydraulic conductivity of discrete soil samples and mass balance calculations. The Board recognizes that natural processes might modify hydraulic conductivity after bottom soils have been inundated. However, these processes are poorly understood at the present time. It is therefore difficult to predict how much the hydraulic conductivity may change and the time it might take to do so. High initial hydraulic conductivity might allow enough seepage to severely degrade underlying and adjacent ground water before natural processes could significantly reduce pond seepage.

The California Central Valley Regional Water Quality Control Board views the ponds as an interim method of disposal of subsurface agricultural drainage. The current regulatory measures for existing and new ponds will not protect the underlying ground water indefinitely. Until a long-term solution to the San Joaquin Valley drainage problem is implemented, the Evaporation Pond Investigation will continue to research and assist in developing plans and policies for operation of evaporation ponds to ensure protection of the environment.

References

California Department of Water Resources (1988), "Agricultural Drainage Evaporation Ponds in the San Joaquin Valley, Progress of the Investigation," Memorandum Report.

California Central Valley Regional Water Quality Control Board (1988), "Water and Sediment Quality in Evaporation Basins Used for Disposal of Agricultural Subsurface Drainage Water in the San Joaquin Valley, California."

California Central Valley Regional Water Quality Control Board (1993a), "Agricultural Subsurface Drainage Evaporation Basins, Tulare Lake Basin," Staff Report.

California Central Valley Regional Water Quality Control Board (1993b), "Order No. 93-136, Waste Discharge Requirements for Tulare Lake Drainage District; North, Hacienda and South Evaporation Basins; Kings and Kern Counties."

California Central Valley Regional Water Quality Control Board (1993c), "Order No. 93-138, Waste Discharge Requirements for Westlake Farms, Inc.; North and South Basins; Kings County."

McCullough-Sanden, B. (1986), "Evaluation of Seepage in an On-Farm Evaporation Pond," Master's thesis, University of California.

Impact of Farming Systems on Water Quality in Iowa[1]

Jerry L. Hatfield   James L. Baker   Phillip J. Soenksen[2]

Non-member   Non-member   Non-member

Introduction

Management practices influence the occurrence and movement of herbicides and nitrate-nitrogen in soil and water resources, yet the controlling processes are not clearly understood. A regional scale effort began in 1990 to evaluate the effect of farming practices on water quality. As detailed by Onstad et al. (1991), the Management Systems Evaluation Area (MSEA) program has two main goals: 1) to evaluate the distribution of agricultural chemicals in water resources and identify the processes and factors that affect distribution, and 2) to develop alternative and innovative agricultural management systems that enhance and

Contribution from the USDA-Agricultural Research Service, Iowa Agricultural Experiment Station, and United States Geological Survey-Water Resources Division.

[2]Laboratory Director, USDA-ARS, National Soil Tilth Laboratory, Ames, Iowa; Professor, Department of Agricultural and Biosystems Engineering, Iowa State University, Ames, Iowa; and Hydrologist, Water Resources Division, United States Geological Survey, Iowa City, Iowa, respectively.

247

protect water quality. Quantitative approaches based on environmental quality standards have been unused in the development of improved management practices. The MSEA program is a regional effort and covers 10 sites in the Midwest with research centers in Iowa, Minnesota, Missouri, Nebraska, and Ohio. Quantitative approaches can be used to develop an understanding of the different processes involved within the soil that may be influenced by changes in management practices. This report describes the research objectives and sites of the MSEA project within Iowa and highlights some of the accomplishments to date.

Iowa has been recognized for its efforts in the water quality area for a number of years. Two of the first multiagency projects conducted within the state were centered in Tama County and in the Big Springs area as described by Johnson and Baker (1984), Hallberg and Hoyer (1982), and Hallberg et al. (1984). These projects raised the awareness of farmers and the public about the potential links between agricultural practices and effects on water quality. Hallberg (1989) reported that nitrate and herbicides have been found in many of the Iowa aquifers and suggested that the detections were related to the increased usage of agrichemicals. The Iowa Well Water Survey (Iowa Department of Natural Resources, 1990) provided a thorough analysis of the quality of ground water and suggested complex relationships with agricultural practices, depth of the well, geology of the area, hydrology, and soils. This study found herbicides in many of the wells throughout Iowa, and the detection was related to geological structure and patterns of herbicide use. Occurrences were more frequent in shallow wells than in deeper wells. Baker et al. (1985) found nitrate-nitrogen levels in excess of 10 mg/L in ground water from 25% of 170 farm wells in Iowa.

Goolsby and Battaglin (1993) analyzed data from surface water samples from the Midwest and found that concentrations and mass transport of herbicides follow an annual cycle. In their reconnaissance study, several of the drainage basins in the Mississippi River watershed were sampled beginning in 1989. They found that less than 3 percent of the herbicide mass applied to cropland was transported into streams; however, this mass was sufficient to cause atrazine concentrations to exceed 3 ug/L in the Mississippi River for short periods of time. The peak herbicide concentrations were found in storm runoff in May, June, and July with some detections throughout the year. Concentrations were related to the amount applied. Nitrate-nitrogen concentrations throughout the year exhibited a different pattern than herbicides with the highest concentrations in the winter and spring and lowest during the summer. Kolpin et al. (1993) found detectable herbicide or atrazine metabolite in 28.4% of the 303 Midwestern wells sampled in 1991. They also found that none of the wells had concentrations that exceeded the maximum contamination levels for drinking water.

MSEA Objectives

The objectives of the Iowa MSEA project were developed to cover a range of temporal and spatial scales and to incorporate the expertise of researchers representing a wide range of disciplines. The objectives are:

1. Quantify the physical, chemical, and biological factors affecting the transport and fate of agricultural chemicals.

2. Determine the effects of crop, tillage, and chemical management practices on the quality and quantity of surface runoff, subsurface drainage, and ground-water recharge.

3. Integrate information from Objectives 1 and 2 with data about soil, atmospheric, geologic, and hydrologic processes to assess the impact of these factors on surface and ground water quality.

4. Evaluate the socioeconomic impacts of current and developmental management practices that may emerge as most effective from Objectives 1, 2, and 3.

5. Develop an education and technology transfer program that would quickly disseminate information to the users.

Research Locations and Procedures

Research efforts are underway in three different settings at four major sites: thick loess - Deep Loess Research Station near Treynor (southwestern Iowa), thin till over bedrock - Northeast Research Center near Nashua (northeastern Iowa), and thick till - Till Hydrology site and Walnut Creek watershed near Ames (central Iowa). These sites are characteristic of 35 percent of the land area of the state. Each site adds a particular research history to the project and represents typical farming practices for the area. In the development of the Iowa MSEA project, existing sites and ongoing studies were used, where possible, to fulfill the objectives of the MSEA project and assess the impact of current farming practices. Ongoing studies were used so that observations of the fate and transport of agrichemicals in the water and soil would be representative of that practice. This design also permitted the incorporation of the current understanding about the response of a particular site to the different practices. Atrazine, alachlor, metribuzin, metolachlor, and nitrate-nitrogen are measured at all sites in both water and soil matrices.

Treynor - Thick Loess

Deep loess soils overlie rolling glacial till in western Iowa in Major Land Resource Area (MLRA) 107. The topography consists of narrow, gently sloping ridges, steep side slopes, and well-defined alluvial valleys. A typical cross section of the loess hills is shown in Figure 1. Saturated zones within the loess and till serve as water sources for domestic uses. Nitrate-nitrogen concentrations in shallow wells (less than 10 m) within the loess often exceed 10 mg/L. Wells within the glacial till, however, exhibit better water quality. Given the steep slopes, these soils are subject to erosion and large amounts of surface runoff. Streams in the alluvial valleys are fed by both surface runoff from the upland areas and seepage flow from the loess-till interface. These loess soils are in the Marshall-Monona-Ida-Napier series; they are well-drained and have high water holding capacities.

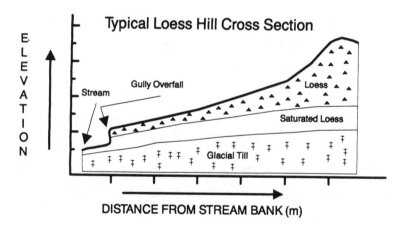

Figure 1. Cross-section of the deep loess soils and the underlying glacial till typical of the deep loess area.

Rainfall at the site averages 820 mm per year, concentrated during the spring and summer. Significant surface runoff events can occur in the spring when large intense storms occur before the crop is established, and when snow melt occurs while underlying soil is still frozen.

Four field-sized watersheds which range from 30 to 61 ha are being studied at Treynor. Corn is grown on all watersheds each year. On two watersheds, the primary tillage is deep tandem disking with spring broadcast application of herbicides and nitrogen applied as anhydrous ammonia. Ridge tillage is used on the other two watersheds, one of which has a series of parallel terraces drained by an underground pipe system via surface inlets. Research studies have been ongoing since 1964 and are described by Hatfield et al. (1993).

In each of the watersheds, the amount and quality of flow, including base or seepage flow from the interface of the loess and glacial till, are determined at the watershed outlet. During surface runoff events, sediment samples are also collected. In 1991 additional monitoring stations were established in each watershed to collect samples at the edge of the field to permit a separate collection of the surface runoff. Recording raingages are located throughout the watersheds to determine rainfall amounts, durations, and intensities. An automated weather station is located near each watershed to record meteorological and soil temperature data. Crop yield information and corn stalk data are collected to quantify crop nitrogen uptake.

Soil samples are collected to a depth of 1.2 m at various locations throughout each watershed eight times per year to quantify the soil profile concentrations of metolachlor and atrazine. These samples are collected before spring application, immediately after application, at 2, 4, 6, 8, and 14 weeks after planting, and after harvest.

In the fall of 1991, four well nests were installed in the ridge tillage watershed to allow measurement of chemicals in the soil water extracted from the wells and volatilized material collected in the soil air removed from different depths. These well nests are arranged to measure a transect within the field from the upper slopes down to the valley. Suction lysimeters were placed under the lower slopes and valley area of this watershed to collect lateral flow of water moving within the soil profile. These collection systems are sampled monthly throughout the frost-free period.

Nashua - Thin-Till

This site is in MLRA 104 with topography ranging from level to gently rolling. Soils at the Nashua site are in the Kenyon-Floyd-Clyde association. Although the soils generally are thin, they are moderately to poorly drained requiring the extensive use of tile drains. The geology is Pre-Illinoian glacial till which overlies Devonian age carbonate bedrock. A geologic log, compiled from well cuttings obtained during the installation of a nest of piezometers, is shown in Figure 2. Tile lines, which divert excess water from the soil to streams, offer a unique way to sample root zone water under different agricultural practices. Annual rainfall in this area averages 850 mm.

The same tillage, crop, fertilizer, and chemical practices have been used at the Nashua site since 1975 as described in Hatfield et al. (1993). There are 36, 0.43

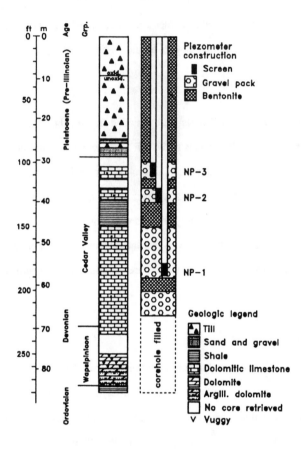

Figure 2. Geologic log of surficial and bedrock deposits in the Nashua area (detailed geologic log provided by W. Simpkins, 1993).

ha plots with parallel tile drains down the center and alongside of each plot at a depth of 1.2 m. The center tiles drain into a sump pump system that measures and samples drainage water from each plot. Flumes with flow meters to measure surface runoff and automatic water quality samplers were installed in early 1992 on four plots; tillage practices were different on each plot. Four tillage practices and two crop rotations are being studied at Nashua. Each crop of the corn-soybean rotation is grown each year. In 1991, piezometers were placed around each plot to allow measurement of hydraulic head changes below the tile drains. An irrigation system was installed to provide some control of the soil water availability to particular plots and practices.

The tillage, crop, fertilizer, and chemical practices at Nashua were changed in 1993 to allow testing of alternative practices compared to existing practices for their effect on water quality. These practices are described by Hatfield et al. (1993).

Ames - Thick-Till

The topography of MLRA 103 can be characterized as level to gently sloping to irregular with poor surface drainage features, including the presence of prairie potholes. Potholes, with no natural surface outflow, are present in many areas of this region. Tile drains were installed in many places in the early 1900's to remove water from the potholes and direct it to nearby ditches and streams. Flow from the tiles into the streams usually begins in the early spring as the soil begins to thaw and continues throughout the summer and fall. If rainfall is below normal, tile flow will cease sooner than usual. Tile drains generally range in size from 10 to 90 cm in diameter; the larger sizes serve as main branches while the smaller sizes drain individual potholes or fields. Soils within the study areas near Ames are in the Clarion-Nicollet-Webster association. The Clarion soils are well-drained and moderately permeable on the upland sites, while the adjacent Nicollet soils are poorly drained with moderate permeability. Average annual rainfall is 840 mm with most of the rainfall occurring in the spring and summer months.

The geology of the area under the Ames study sites is complex, as shown in Figure 3, with extensive lateral and vertical variation in deposits. Surficial deposits of Wisconsinin till overlie Pre-Illinoian till which contains a laterally extensive gravel lens. Beneath these unconsolidated materials lie Pennsylvanian shales and sandstones which, in turn, are underlain by Mississippian carbonate rocks. Deep piezometers were installed in 1990 at the Till-Hydrology Site to collect water samples from the geologic units as shown in Figure 3. These piezometers have been used to collect water samples for ^3H and ^{13}C analyses to provide a measure of the age of water within each geologic unit.

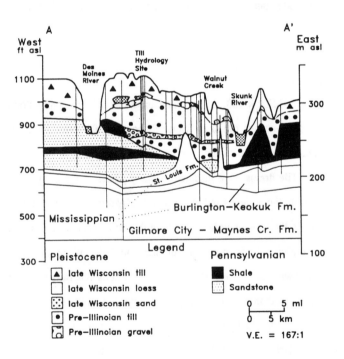

Figure 3. Cross-section of the geologic units in the Walnut Creek area near Ames (detailed cross-section provided by W. Simpkins, 1993).

At the Till-Hydrology Site, eight, 0.43 ha plots have been used since 1985 to measure the effect of tillage; no-till versus moldboard-plow; fertilizer management, single versus split application of nitrogen; and herbicides on tile drainage and ground water quality for continuous corn. Soil samples have been collected from these plots to measure the profile distribution of atrazine and its metabolites and to determine the possible effect of preferential flow on the movement of chemicals through the soil profile.

Walnut Creek watershed was selected as an area typical of MLRA 103 and capable of fulfilling the MSEA objectives at a scale not being studied at the time. The topography of this 5600 ha watershed is generally flat with numerous potholes in the upper reaches, level to gently rolling in the middle reaches, and fairly steep near the stream in the lower reaches. Tile drains are used extensively in the upper reaches, but are used in few places in the lower reaches. Specific fields were selected in the watershed to represent different farming practices. Changes were made in other fields to provide a field-scale comparison of farming practices. All field operations are conducted by the farmers, and measuring equipment is placed to provide minimal disturbance to farming operations.

Extensive monitoring systems were installed throughout the watershed beginning in 1990 (Hatfield et al., 1994). Wells, arranged to measure different intervals to depths of 10 m, were installed in and around individual fields and one subbasin. Ground water levels are measured biweekly and water quality samples are collected monthly. Tile flow (two fields and three subbasins), surface runoff (two fields and one subbasin), and streamflow (three main stream and one tributary) are measured and recorded electronically on five-minute intervals at numerous sites. Water quality samples are collected automatically, based on real-time flow data, at each of these sites as are sediment samples at surface runoff and streamflow sites. Manual water quality samples are collected weekly at these and numerous other tileflow and streamflow sites, including the South Skunk River and several sites in adjacent watersheds.

Two meteorological stations were placed in the watershed, one at the upper end and the other near the middle. These stations record air temperature, humidity, solar radiation, wind speed and direction, rainfall, and soil temperature. At each of these stations, a wet/dry precipitation sampler collects rainfall for water quality analyses. Throughout the watershed, 20 tipping bucket raingages have been placed to determine rainfall amounts, durations, and intensities. Six Bowen ratio systems are placed in different fields to measure the energy balance throughout the year and the latent and sensible heat flux changes due to different surface management practices.

A geographical information system (GIS) database has been developed for Walnut Creek. Attributes include: topography, soils, tile drains, field boundaries, land use, chemical use, farmsteads, roads, and streams. These data serve as inputs into the chemical transport, streamflow and ground water models that will allow different practices to be compared across the landscape. All of the data which are collected within Walnut Creek are being placed into a relational data base with the use of SAS procedures (SAS, 1992). Several field and watershed scale models are being parameterized and evaluated based on data from Walnut Creek. These include the Precipitation Runoff Modeling System (PRMS), GLEAMS,

Field Soil Water Balance Model with PRZM, and Root Zone Water Quality Model (RZWQM) as initial candidates for watershed assessment and design.

Research Results

At Walnut Creek watershed, nitrate-nitrogen concentrations in the weekly tileflow and streamflow samples show a large variation throughout the year with concentrations in excess of 15 mg/L during the spring and summer (Figure 4).

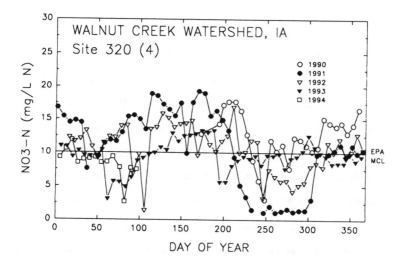

Figure 4. Temporal sequence of nitrate-nitrogen concentration in the stream flow at one site within Walnut Creek watershed.

During the late summer and early fall there is typically a decrease below the MCL of 10 mg/L. Even though the years have been different in rainfall amounts and temporal distribution of rainfall, there is a large amount of uniformity among the years. Event samples of streamflow show that short-term nitrate-nitrogen concentrations remain fairly steady during periods of base flow, but during stormflow, concentrations decrease while loads increase. Samples collected from shallow wells surrounding fields show nitrate-nitrogen concentrations less than 5 mg/L and decreasing with depth below 3 m. Soil profile samples show nitrate-nitrogen concentrations above 30 mg/L in the upper 30 cm decreasing to less than 10 mg/L below this depth. Efforts are underway at fields within Walnut Creek to determine if the temporal distributions of nitrate are related to the management practice.

At Treynor the nitrate concentrations in the seepage flow are related to the tillage practices. Kramer et al. (1990) found that the nitrate loss was greater under the ridge tillage watersheds compared to the conventional tillage because surface runoff was reduced resulting in a greater leaching through the soil profile. Over 20% of the applied nitrate was lost through leaching. At the thin-till site, Kanwar et al. (1991) found a similar pattern with increased drainage through the ridge tillage and no-till plots and a greater nitrate loading rate in the tile drainage. In their study the concentrations of nitrate were reduced compared to the moldboard plow and chisel plow. Several studies are underway to determine the mechanisms that influence the nitrate patterns under different tillage practices. In a companion study conducted at Walnut Creek and Treynor, Hatfield and Prueger (Unpublished data) have found that the change in tillage practices influences the soil water evaporation rate in the early season. Current tillage practices do not use this stored water during the season except in the case of limited soil water. In most years the surplus water moves through the soil profile and carries the nitrate below the root zone. Developing improved water management in crop rotations and during the growing season in the humid regions may be a viable tool for improving water quality.

Baker and Kanwar (Unpublished data) have found that split application of nitrogen fertilizer reduced the nitrate-nitrogen concentration in the tile drainage water at the Till-Hydrology site. Nitrate-nitrogen concentrations from a single application of 175 kg/ha of N exceeded 15 mg/L during the spring and summer while from the split applications totaling 125 kg/ha of N, concentrations were less than 10 mg/L. Field scale studies are currently underway in Walnut Creek watershed to evaluate split applications of N compared to a single fall application of anhydrous ammonia and to a swine manure application.

Concentrations of atrazine, metolachlor, metribuzin, and alachlor in the streamflow and tile drainage of Walnut Creek watershed generally are below the MCL during baseflow but increase and decrease rapidly during stormflow as shown in Figure 5 for atrazine during a storm in 1991. Peak concentrations and total loads during stormflow events decrease with time since application. For a 2540 ha subwatershed of Walnut Creek during the period from April 1991 through September 1992, Soenksen et al. (1993) found chemical yields in streamflow to be 58.4 kg/ha for nitrate-nitrogen, 4.5 g/ha for atrazine, and 7.3 g/ha for metolachlor. The atrazine lost in the streamflow was 0.006 of the total applied per ha in this subbasin while for metolachlor this represented 0.009 of the amount applied. Soil profile samples revealed that the highest herbicide concentrations are in the upper 30 cm and are less than 1 ug/L below this depth. Atrazine has been found in the soil profile in fields where there has been no application for over eight years. Studies are underway at these fields to examine

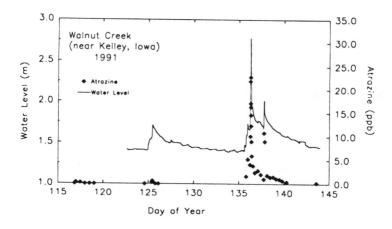

Figure 5. Stream flow height and atrazine concentration for a runoff event in 1991 within Walnut Creek watershed.

the metabolite concentrations within the soil profile as a measure of the degradation and transformation processes within the soil. Samples collected from the shallow wells revealed that there were detectable amounts of atrazine and metolachlor in some cases; however, these detections were not consistent across sampling events. Samples collected from deep wells (below 50 m) within Walnut Creek have shown no detections of these herbicides and nitrate-nitrogen concentrations below 2 mg/L.

Analyses are underway to determine the attributes of farming systems most closely coupled with herbicide and nitrate transport. These studies are conducted in both the laboratory and field to provide a range of scales under which variables are controlled. The sociological and economic studies conducted as part of Objective 4 focus on the adoption of technology and the dissemination of results to the user. Surveys are being conducted to determine how farmers view the water quality problem and how they proceed in the collection and assimilation of information before making management decisions.

References

Baker, J.L., R.S. Kanwar, and T.A. Austin. 1985. Impact of the use of agricultural drainage wells on ground water quality. J. Soil and Water Cons. 40:516-520.

Goolsby, D.A. and W.A. Battaglin. 1993. Occurrence, distribution, and transport of agricultural chemicals in surface waters of the Midwestern United States. In Selected Papers on Agricultural Chemicals in Water Resources of the Midcontinental United States. U.S. Geological Survey Open File Report 93-418. Denver, CO. pp. 1-25.

Hallberg, G.R. and B.E. Hoyer. 1982. Sinkholes, hydrogeology, and groundwater quality in northeast Iowa. Iowa Geological Survey, Open-File Report 82-3. 120 p.

Hallberg, G.R., R.D. Libra, E.A. Bettis III, and B.E. Hoyer. 1984. Hydrogeologic and water quality investigations in the Big Springs basin, Clayton County, Iowa. 1983 Water-Year. Iowa Geological Survey, Open-File Report 84-4. 231 p.

Hallberg, G.R. 1989. Pesticide pollution of groundwater in the humid United States. Agric. Ecosystems Environ. 26:299-367.

Hatfield, J.L., J.L. Baker, P.J. Soenksen, and R.R. Swank. 1993. Combined agriculture (MSEA) and ecology (MASTER) project on water quality in Iowa. In Agricultural Research to Protect Water Quality. Soil and Water Conservation Society, Ankeny, IA. pp. 48-54.

Hatfield, J.L., P.J. Soenksen, R.C. Buchmiller, and R.R. Swank. 1994. Walnut Creek Watershed: Experiment Design and Sampling Strategies. National Soil Tilth Laboratory Technical Report 93-1. (In Press.)

Iowa Department of Natural Resources. 1990. The Iowa State-Wide Rural Well-Water Survey. Water Quality Data: Initial Analysis. Tech. Information Series Number 19. 142 p.

Johnson, H.P. and J.L. Baker. 1984. Field-to-stream transport of agricultural chemicals and sediment in an Iowa watershed: Part II. Data base for model testing (1979-1980). Report No. EPA-600/S3-84-055, Environmental Research Laboratory, Athens, GA.

Kanwar, R.S., D. Stoltenberg, R. Pfeiffer, D.L. Karlen, T.S. Colvin, and M. Honeyman. 1991. Long-term effects of tillage and crop rotation on the leaching of nitrate and pesticides to shallow groundwater. In W.F. Ritter (ed.) Proc. of ASCE Irrigation and Drainage Conference. ASCE. New York. pp. 604-610.

Kolpin, D.W., D.A. Goolsby, D.S. Aga, J.L. Iverson, and E.M. Thurman. 1993. Pesticides in near-surface aquifers: Results of the Midcontinental United States Ground-Water Reconnaissance, 1991-92. In Selected Papers on Agricultural Chemicals in Water Resources of the Midcontinental United States. U.S. Geological Survey Open File Report 93-418. Denver, CO. pp. 64-74.

Kramer, L.A., E.E. Alberts, A.T. Hjelmfelt, and M.R. Gebhardt. 1990. Effect of soil conservation systems on groundwater nitrate levels from three corn-cropped watersheds in Southwest Iowa. Proc. of Agricultural Impacts on Groundwater Quality. Am. Soc. Groundwater Scientists and Engineers. pp. 175-189.

Onstad, C.A., M.R. Burkart, and G.D. Bubenzer. 1991. Agricultural research to improve water quality. J. Soil and Water Conserv. 46:184-188.

SAS. 1992. SAS Users Guide. SAS Institute Inc., Cary, NC 27512-8000.

Soenksen, P.J., J.L. Hatfield, and D.J. Schmitz. 1993. Chemical loads of nitrate, atrazine, and metolachlor in Walnut Creek watershed near Ames, Iowa, 1991-92. Proc. of the Toxic Substances Hydrology Program, Technical Meeting, Denver, CO, Sept. 20-24, 1993.

Impact of Nitrogen and Water Management on Ground Water Quality

J.S. Schepers, M.G. Moravek, R. Bishop, and S. Johnson[1]

Introduction

Nitrate (NO_3) contamination of shallow ground water aquifers that are used for drinking water is a serious problem in many parts of the world. Trends in the occurrence of nitrate-nitrogen (NO_3-N) in ground water at levels above the recommended safe drinking water standard of 10 mg NO_3-N/L are not encouraging. Areas with histories of intensive cultivation have an increased potential for nitrate leaching (Schepers et al., 1986; Spalding and Kitchen, 1988). Sources of NO_3 moving through soil and toward the ground water are difficult to identify unless isotopic N tracers are used. Therefore, targeting production practices that likely degrade ground water quality must be done while recognizing this uncertainty. Nevertheless, thorough examination of cropping systems, soil characteristics, and climatic

[1] Soil Scientist, USDA-Agricultural Research Service and Professor of Agronomy, Univ. of Nebraska, Lincoln, NE 68583-0915, Assistant Manager and Manager, Central Platte Natural Resource District, 215 North Kaufman Ave., Grand Island, NE 68803, and undergraduate student, University of Nebraska, Lincoln, NE.

factors can identify production practices and physical variables that should be targeted for improved management.

Although it is possible to identify variables and production practices that could be modified to reduce the potential for NO_3 leaching, changes in production systems usually are not made without tradeoffs (Watts and Martin, 1981). These modifications may involve additional labor, shifting of operations to other times, additional technical expertise, or increased expenses. Any or all modifications to cropping systems involve uncertainties that can affect profitability. Therefore, changes in cropping systems for the purpose of protecting water quality are likely to occur gradually unless they are motivated by legislation or induced by financial incentives.

Nonpoint source contamination of ground water represents a unique challenge for society because it is difficult to impose regulations without differentially impacting current operations of various individuals or groups of producers. The purpose of this paper is to examine the ground water protection program initiated in 1987 by the Central Platte Natural Resources District (CPNRD) in Central Nebraska and evaluate production and water quality data from fields within a 200,000-ha area that is dominated by irrigated corn production.

Nitrogen and Water Management

Producers in the CPNRD Phase II Area (average ground water NO_3-N concentrations between 12.6 and 20.0 mg/L) are required to sample the soil in each field to a depth of 0.9 m prior to planting and to determine the NO_3-N concentration in irrigation water. This information, along with an estimate of expected yield, is used to make fertilizer N recommendations. Soil residual N, water amount and NO_3-N concentration, and production records for individual fields must be provided to the CPNRD each fall. Schepers et al., (1991a) summarized these data for 1988 and then updated the records to include the 1989 growing season (Schepers et al., 1991b). This report summarizes the producer's N and water management activities from 1988 thru the 1992 growing season.

Perceived causes of NO_3 contamination of ground water focus on excess application of N fertilizer. Other factors, such as overly optimistic yield expectations, also tend to increase fertilizer N recommendations. Poor water management will also enhance NO_3 leaching. Data by Ferguson et al. (1990) from a limited number of irrigation wells in the Platte River Valley showed annual application amounts ranging from 0 to 135 cm and averaging 38 cm from 1979 to 1983. Unfortunately, few irrigation wells utilize flow meters. Therefore, estimates of seasonal water application amounts are not reliable. Complete data sets from over 3000 fields in 1988 thru 1992 were grouped and summarized on the basis of yield goals compared to yields attained and on differences between the amounts of fertilizer N recommended and applied.

Grain yields over the five years averaged 9.90 Mg/ha, which averaged 10% less than the yield expectations (Table 1). Distribution of relative yields followed the same general shape for all years (Figure

1). This optimism translates into approximately 22 kg/ha excess N in terms of fertilizer recommended.

Table 1. Production characteristics for the CPNRD Phase II Ground Water Protection Area*.

Year	Area	N recommended	N Applied	Dev. from recommended N	Yield	Relative yield
	(1000 ha)	(kg/ha)	(kg/ha)	(kg/ha)	(Mg/ha)	
1988	84.1	116	164	47	9.99	.91
1989	96.9	124	156	31	9.92	.90
1990	107.8	114	143	30	9.86	.90
1991	111.7	126	156	28	9.91	.90
1992	102.0	134	156	20	9.82	.89

* Weighted by area of each field.

Producers realize that climate influences grain yield, so there is a tendency to hope for favorable growing conditions and fertilize for the associated higher yields. Many producers do not realize that more favorable temperature and soil water conditions associated with higher levels of production also result in higher levels of N mineralization (ie., microbial decomposition of organic matter). As such, some natural processes such as crop N uptake and soil microbial activity tend to offset one another within reasonable limits.

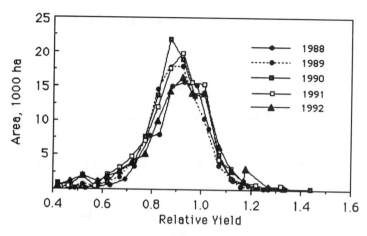

Figure 1. Distribution of irrigated corn production relative to the fraction of the yield goal attained by producers.

Fertilizer N applications in excess of CPNRD fertilizer guidelines, which were developed by the University of Nebraska, are an obvious source of leachable N and possible cause for ground water contamination. Fertilizer applications within 22 kg N/ha of the recommended rate are probably within the accuracy of application equipment. In the first year of the program, fields accounting for 58% of the area received fertilizer N applications in excess of CPNRD recommendations (Table 2). By 1992, corn fields receiving fertilizer N applications in excess of CPNRD recommendations declined to 43% of the area. Most of this decline in over application was attributed to producers who had previously applied at least 67 kg N/ha more than recommended. Throughout the study, ~10% of the area received less N fertilizer than recommended (<22 kg N/ha). Reasons given for applying less N fertilizer than recommended generally relate to past experiences.

Table 2.   Proportion of the CPNRD Phase II Ground Water Protection Area receiving various amounts of fertilizer N.

Year	Fertilizer N applied minus N recommended (kg/ha)					
	<-67	-67 to -22	-22 to 22	22 to 67	67 to 112	>112
	-----------------------------------(% of area*)----------------------------------					
1988	1.4	8.8	31.9	31.7	15.2	10.9
1989	1.0	9.3	35.7	29.1	17.9	7.0
1990	1.2	7.7	37.9	31.1	17.6	4.6
1991	0.8	7.4	42.9	29.5	14.7	4.7
1992	0.8	10.4	45.7	27.9	12.7	2.4

* Area under corn production for each year is noted in Table 1.

Reasons why producers apply more fertilizer N than recommended include 1) compensation for potential leaching and denitrification losses, 2) uncertainty over how corn hybrids respond to even a slight N deficiency, 3) awareness that grain yields generally increase with increasing fertilizer N rates up to some poorly defined upper limit, 4) compensation for field variability and a concern over even a part of the field showing an N deficiency, and 5) concern that other N sources such as residual soil N, mineralized N, and NO_3 in irrigation water may not be as available to the crop as is fertilizer N. While it is not possible to assess the merits of the individual concerns above, the data support the premise that producers tend to apply enough fertilizer N to compensate for N removed in grain (Figure 2). In general, many producers who were advised to apply little or no fertilizer N still applied generous amounts. In contrast, others who were advised to apply more fertilizer N than what they thought was needed according to a general "rule of thumb" tended to apply less than the recommended amount.

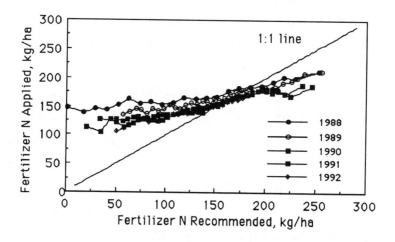

Figure 2. Fertilizer N application rates as related to recommended amounts for irrigated corn production.

Producers with small fertilizer N recommendations in 1988 questioned the reliability of the N management guidelines because of the perceived need for early season N to aid in plant establishment.

Because producers in the area are inclined to apply a small amount of N fertilizer, even though none is recommended, the CPNRD recommended a minimal application of 56 kg/ha to all fields beginning in 1989. As such, an average of 15.8% of the area received the minimal fertilizer N recommendation from 1989 through 1992. This group of producers applied an average of 133 kg/ha of N fertilizer, which was 77 kg N/ha more than recommended. About 46% of the area that received the minimal N recommendation of 56 kg/ha would have otherwise received a "zero" fertilizer N recommendation had the minimal recommendation not been implemented. Average yield for producers receiving the minimal fertilizer N recommendation was 9.55 Mg/ha compared to 9.88 Mg/ha for the entire area. While this yield difference is small and probably not significant, fields receiving the minimal fertilizer N recommendation had an average ground water NO_3-N concentration of 24.7 mg/L. There should be little doubt that the high levels of residual soil N for this group of fields and overfertilization contribute significantly to ground water contamination. Even greater potential for NO_3 leaching is associated with seed corn production (averaged 1.9% of the area), where fertilizer N application rates averaged 122 kg/ha while grain removal is estimated to be 63 kg/ha.

Targeting specific groups of producers for improved N and water management has merit if the socio-economic profile of these producers can be developed. Trends in overfertilization in the area were consistent over time (Figure 3) and has probably occurred for a number of years. The effect of overfertilization on residual soil N levels after harvest has not been well documented for specific fields in the CPNRD project area.

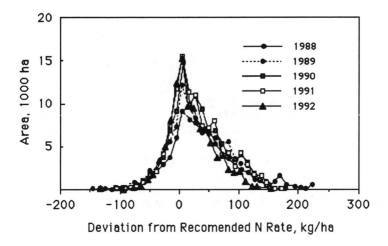

Figure 3. Distribution of irrigated corn production for various amounts of N fertilizer relative to recommended amounts.

Traditionally, producers do not make drastic changes in N management practices from year to year, so it is probably reasonable to assume that fields receiving excess N fertilizer should also have higher levels of residual soil N (Figure 4). This line of reasoning is supported by the observation that fields receiving recommended fertilizer N rates had consistently higher yields than those receiving less than recommended N rates but had similar levels of residual soil N.

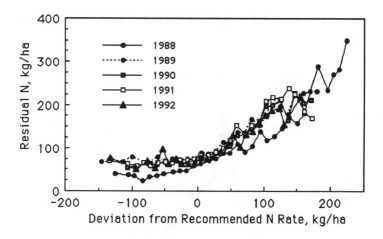

Figure 4. Residual soil N for fields receiving various amounts of N fertilizer relative to recommended amounts.

Concentrations of NO_3-N in irrigation water for the various categories of fertilizer application followed similar trends for each of the five years (Figure 5). Scatter in the data at low and high fertilizer N rates are attributed to the limited number of fields in these categories. The general increase in ground water NO_3-N concentration is expected because of the pertetual over application of N fertilizer. It has taken several decades for the extensive sand and gravel aquifer (15 to 20 m thick) to reach current levels of NO_3-N. As such, reducing NO_3-N concentrations in the aquifer will be slow even if all sources of contamination are eliminated.

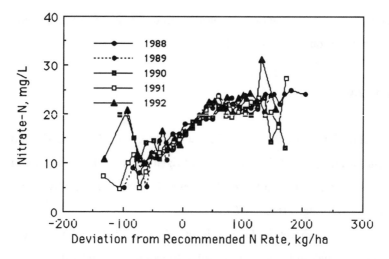

Figure 5. Nitrate-N concentration in irrigation wells for fields receiving various amounts of N fertilizer relative to recommended amounts.

Preliminary data indicate that some producers apply several times the amount of irrigation water required to satisfy crop needs because of inexpensive pumping costs and difficulties achieving adequate water application to difficult to irrigate areas of the field, which enhances NO_3 leaching and ground water contamination.

Information generated from the CPNRD data base was used to develop N and water management practices for Nebraska's Management Systems Evaluation Area (MSEA) project. The MSEA project, which was initiated in 1990, is one of five such projects in the United States as part of the President's Water Quality Initiative. Special emphasis focuses on developing cropping systems that integrate N and water management practices to protect ground water resources while sustaining productivity and profitability.

Literature Cited:

Ferguson, R.B., Eisenhauer, D.E., Bockstadter, T.L., Krull, D.H., and Buttermore, G. 1990. Water and nitrogen management practices in the Central Platte Valley of Nebraska. J. Irr.and Drain. Div., Am. Soc. Civil Eng. 116:557-565.

Schepers, J. S., Frank, K. D., and Bourg, C. 1986. Effect of yield goal and residual soil nitrogen considerations on nitrogen fertilizer for irrigated maize in Nebraska. J. Fert. Issues 3:133-139.

Schepers, J.S., Moravek, M.G., Alberts, E.E., and Frank, K.D. 1991a. Cumulative effects of fertilizer and water management on nitrate leaching and ground water quality. J. Environ. Qual. 20:12-16.

Schepers, J.S., Moravek, M.G., and Bishop, R. 1991b. Impact of nitrogen and water management on ground water quality. Irrigation and Drainage. p. 641-647. W.F. Ritter (ed.) Am. Soc. of Civil Eng. New York, N.Y.

Spalding, R.F., and Kitchen, L.A. 1988 Nitrate in the intermediate vadose zone beneath irrigated cropland. Ground Water Management Research. Spring 1988. pp. 89-95.

Watts, D.G., and Martin, D.L. 1981. Effects of water and nitrogen management on nitrate leaching loss from sands. Trans. Am. Soc. Agric. Eng. 24(4):911-916.

SUMMARY
Wendy L. Cohen, M.ASCE[1]

An awareness of the importance of ground water protection came about in the 1970's. The federal government adopted the first ambient water quality standards in 1978, with many more standards adopted throughout the 1980's. However, there is no comprehensive federal legislation dealing with ground water protection. Federal policies have been aimed primarily at surface water protection where point source pollution control is most important. More recently, the emphasis is shifting to address nonpoint source pollution control through the use of best management practices.

As this compilation of papers illustrates, many different approaches are used around the country to protect ground water quality. Strategies often target specific ground water systems for protection based on the presence of water supply wells, the population dependent on the ground water, and the vulnerability and yield of the aquifer. Using antidegradation policies, some states seek to protect the existing and potential future beneficial uses of ground water by minimizing the anthropogenic effects on ground water quality. Computer modeling of ground water flow and contaminant transport can provide an understanding of the hydrogeology which is critical to any comprehensive protection strategy. For nonpoint source pollution, best management practices are used as the most effective and practical control measures

Following are some specific protection alternatives and strategies:

*Land use controls in designated wellhead protection areas (WHPA) where recharge of water supply aquifers occurs.

*Enforcement of numerical standards for ground water.

[1] Senior Water Resources Control Engineer, California Regional Water Quality Control Board, Central Valley Region, 3443 Routier Road, Sacramento, CA 95827

*Regulation of contaminant sources in WHPAs by writing permits and setting effluent limits on discharges of pollutants.

*Coordination of water supply with hazardous waste remedial programs for contaminated wellfields.

*A comprehensive ground water management code designed to reduce long-term overdevelopment and bring overdrafted basins back into hydrologic balance.

*A water supply plan accounting for available water resources, identified problem areas, and potential solutions to those problem areas.

*Examination of agricultural cropping systems, soil characteristics, and climatic factors to find which production practices and physical variables to target for improved management.

*Development of guidelines for design, construction and operation of agricultural evaporation ponds in the most environmentally safe and cost-effective manner.

*Limiting ground water withdrawals to the annual recharge to prevent aquifer overdrafting in formations where a lowering of the water table could result in degradation of the quality of the aquifer.

*Increasing public awareness through the media to reduce ground water pollution at the source.

No specific ground water protection strategy will work in all situations. Each aquifer is unique, and the protection strategy chosen must account for all site specific factors. The message is clear, however, that ground water quality must be protected from all the pollutant sources humans have developed over the years if we are to retain all the benefits of clean and healthy water.

SUBJECT INDEX

Page number refers to the first page of paper

AUTHOR INDEX

Page number refers to the first page of paper